Mathematical Sciences Research Institute
Publications

1

Editors

S.S. Chern
I. Kaplansky
C.C. Moore
I.M. Singer

Mathematical Sciences Research Institute
Publications

Daniel S. Freed
Karen K. Uhlenbeck

Instantons and Four-Manifolds

Second Edition

With 61 Illustrations

Springer-Verlag
New York Berlin Heidelberg London
Paris Tokyo Hong Kong Barcelona

Daniel S. Freed
Department of Mathematics
University of Texas at Austin
Austin, Texas 78712

Karen K. Uhlenbeck
Department of Mathematics
University of Texas at Austin
Austin, Texas 78712

Mathematical Sciences Research Institute
1000 Centennial Drive
Berkeley, California 94720

The Mathematical Sciences Research Institute and the authors wish to acknowledge support from the National Science Foundation.

AMS Subject Classification: 57M40, 57N15, 57R60, 58DXX, 81E99, 81E10

Library of Congress Cataloging in Publication Data
Freed, D. (Daniel)
   Instantons and four-manifolds.—2nd ed.
   (Mathematical Sciences Research Institute
publications ; v. 1)
   Bibliography: p.
   1. Four-manifolds (Topology) 2. Instantons.
I. Uhlenbeck, K. (Karen) II. Title. III. Series.
QA613.2.U35 1984   514'.3   84-8584

Printed on acid-free paper.

Camera-ready copy prepared by the Mathematical Sciences Research Institute.

9 8 7 6 5 4 3 2 1

ISBN-13: 978-1-4613-9705-2   e-ISBN-13: 978-1-4613-9703-8
DOI: 10.1007/ 978-1-4613-9703-8

To our parents

PREFACE TO THE SECOND EDITION

Almost five years have passed since the first edition of *Instantons and Four-Manifolds* appeared. This period has been marked by intense activity leading to a greater understanding of 4-manifolds. One might erroneously conclude that our book is out of date. On the other hand, our treatment of Donaldson's thesis still provides a nice leisurely introduction to the whole subject. This background is essential to a serious study of later developments. Hence this second edition.

We have not attempted any substantial revisions or additions. However, we did add a brief survey of some significant recent advances. The references there provide the reader with a gateway to more recent literature. We have also taken this opportunity to correct some errors and misprints in the first edition. The entire manuscript has been retyped using TᴇX . We hope the result is more aesthetically pleasing.

Many readers have helped us with their encouragement, comments, and corrections, for which we are extremely grateful. The continued support of MSRI is also appreciated. Finally, we thank Springer-Verlag for keeping our book in print.

Dan Freed   Karen Uhlenbeck

Aspen, Colorado

This book is the outcome of a seminar organized by Michael Freedman and Karen Uhlenbeck (the senior author) at the Mathematical Sciences Research Institute in Berkeley during its first few months of existence. Dan Freed (the junior author) was originally appointed as notetaker. The express purpose of the seminar was to go through a proof of Simon Donaldson's Theorem, which had been announced the previous spring. Donaldson proved the nonsmoothability of certain topological four-manifolds; a year earlier Freedman had constructed these manifolds as part of his solution to the four dimensional Poincaré conjecture. The spectacular application of Donaldson's and Freedman's theorems to the existence of fake $\mathbf{R}^4$'s made headlines (insofar as mathematics ever makes headlines). Moreover, Donaldson proved his theorem in topology by studying the solution space of equations—the Yang–Mills equations—which come from ultra-modern physics. The philosophical implications are unavoidable: we mathematicians need physics!

The seminar was initially very well attended. Unfortunately, we found after three months that we had covered most of the published material, but had made little real progress toward giving a complete, detailed proof. After joint work extending over three cities and 3000 miles, this book now provides such a proof. The seminar bogged down in the hard analysis (§6–§9), which also takes up most of Donaldson's paper (in less detail). As we proceeded it became clear to us that the techniques in partial differential equations used in the proof differ strikingly from the geometric and topological material. The latter can be obtained from basic information in standard references and graduate courses, while no standard accessible set of references exists for all the nonlinear analysis. We have attempted to remedy this by including background material in all subjects, but particularly in analysis (meaning nonlinear elliptic partial differential equations).

Specific mathematical debts are owed. First of all, our proof does follow Donaldson in most essential matters, although we provide much greater detail. On the other hand, we give a more concrete proof of the transversality theorems (§3–§4), a slightly different proof of the orientability theorem (§5), and a completely new proof of Taubes' existence theorem using noncompact manifolds (§7). As a byproduct we obtain a new, easy proof of the Removable Singularities Theorem (Appendix D). We are also able to include the newer important technique of Fintushel and Stern (§10). Our second debt is to Michael Freedman. The seminar was his idea. He has also been our Chief Topological Consultant throughout the entire project. Chapter One follows his first lecture, and large parts of the introduction are due to him. Also, we thank the original speakers in the seminar: Michael Freedman, as well as Andreas Floer, Steve Sedlacek, and Andrejs Treibergs. Many other mathematicians contributed ideas, suggestions, and references. We list a few here, extending to them our heartfelt appreciation, and pray that we have not insulted anyone by inadvertent omission: Bob Edwards, Rob Kirby, Richard Lashof, John Lott, Mark Mahowald, Ken Millett, Tom Parker, Mark Ronan, Rick Schoen, Ron Stern, Cliff Taubes, and John Wood. Dan would particularly like to thank his advisor, Iz Singer, for his advice, information, inspiration, and perspective. The bulk of the proofreading was carried out by David Groisser, and the reader will want to join us in praising David and Louis Crane, who have caught several mysterious statements and incomplete proofs.

MSRI has cheerfully and generously provided us many services, from office space and typing on up; their support even covered some airfares. Larry Castro deserves a special award for enduring all of our corrections and revisions—thank God for the word processor! Both Harvard and Northwestern provided short-term office space. Evy Kavaler drew the creative illustrations. Finally, thanks from Dan to Raoul Bott for his warm hospitality and continued guidance.

<table>
<tr><td>Berkeley, California</td><td>Dan Freed</td></tr>
<tr><td>January, 1984</td><td>Karen Uhlenbeck</td></tr>
</table>

SURVEY OF RECENT RESULTS (SECOND EDITION)

A complete survey of developments since the first printing of *Instantons and Four-Manifolds* would be a daunting task for both the authors and the readers. Instead, our modest aim is to indicate some highlights, leaving details and further results to the references listed below. Our readers may amuse themselves at our expense by observing how wrong were our dire predictions in the introduction to the first printing.

While most of the results in this field concern compact 4-manifolds, the existence of an exotic differentiable structure on $\mathbf{R}^4$ remains an astounding fact. The following theorem adds emphasis to this point.

THEOREM 1 (GOMPF–TAUBES [**G1**], [**T1**]). *There is an uncountable collection of fake $\mathbf{R}^4$'s.*

That is, there are uncountably many diffeomorphism classes of 4-manifolds homeomorphic to $\mathbf{R}^4$. In fact, there is a 2-parameter family of such manifolds.

By general principles there can only be countably many differentiable structures on a compact topological 4-manifold. Before discussing the uniqueness (and classification) of differentiable structures on compact manifolds, let us return to existence questions. The main result in this book asserts that if the intersection form of a closed simply connected 4-manifold is definite, then it is diagonalizable. We pointed out some extensions in Chapter 10. The more general results now cover all fundamental groups [**D1**].

Donaldson has extended his work in many other directions. He has given the arguments a more topological foundation and connected them with algebraic geometry. Because of this last connection, we adopt what has become the prevailing convention and study *anti*-self-dual connections. Then over Kähler surfaces these moduli spaces have an algebro-geometric interpretation. Notice that a Kähler manifold has at least one self-dual direction (given by the Kähler form). In the context of this discussion, the main theorem in the text would refer to negative definite forms.

We turn now to topology. There is a natural map

$$\mu : H_2(X) \to H^2(\mathcal{C}_X)$$

which can be described either using a determinant line bundle (for a family of Dirac operators) or purely topologically via slash product with a class in $H^4(\mathcal{C}_X \times X)$. The moduli spaces $\mathcal{M}_k \subset \mathcal{C}_X$ provide "cycles" on which to evaluate cup products of classes in the image of μ. For k large with respect to b_2^+, the boundary of $\mathcal{M}_k$ does not contribute to the pairing with cohomology—$\mathcal{M}_k$ is "closed." Donaldson argues that the evaluations do not depend on the choice of metric, so lead to topological invariants $D_k(X)$ which are polynomials in $H_2(X)$. For k small with respect to b_2^+, explicit properties of $\partial \mathcal{M}_k$ allow computations which yield non-existence results.

THEOREM 3 (DONALDSON [**D1**]). *Let X be a closed oriented smooth 4-manifold with even intersection form. Suppose $b_2^+(X) \leq 2$ and $H_1(X; \mathbf{Z})$ has no 2-torsion if $b_2^+(X) \neq 0$. Then X is homeomorphic to S^4, $S^2 \times S^2$, or to the connected sum $(S^2 \times S^2) \# (S^2 \times S^2)$.*

The invariants (for $k >> b_2^+$) vanish on certain connected sums.

THEOREM 4 (DONALDSON [**D3**]). *Suppose X is a smooth oriented simply connected manifold which is a connected sum $X = X_1 \# X_2$ with $b_2^+(X_i) > 0$. Then $D_k(X) = 0$.*

Our predictions in our first preface concerning the potential value of the theory on complex manifolds of all dimensions have paid off [**D5**], [**UY**], [**Si**]! On a compact Kähler manifold, $\mathcal{M}_k$ is the moduli space of stable bundles of Chern class k. This is the basis of the following theorems.

THEOREM 5 (DONALDSON [**D3**]). *If X is a complex projective surface, then for large k the polynomial $D_k(X)$ does not vanish.*

Combined with Theorem 4 we conclude

COROLLARY 6 (DONALDSON [**D3**]). *If a smooth simply connected projective surface X is decomposable as a smooth connected sum $X = X_1 \# X_2$, then one of X_1 and X_2 has negative definite intersection form.*

There is a family of algebraic surfaces, the *Dolgachev surfaces $S_{p,q}$ (p, q relatively prime)*, each of which is homeomorphic to $\mathbb{CP}^2 \# 9\overline{\mathbb{CP}^2}$.

THEOREM 7 (DONALDSON [**D4**], FRIEDMAN–MORGAN [**FM1**]). *The Dolgachev surfaces $S_{p,q}$, $(p, 1) = 1$ provide an infinite number of inequivalent smooth structures on $\mathbb{CP}^9 \# 9\overline{\mathbb{CP}^2}$.*

An extensive discussion can be found in the survey article by Friedman and Morgan [**FM2**].

The classification story is not finished. We present two conjectures, the first credited to Kas and Kirby, the second to Thom.

CONJECTURE 8 ("THE 11/8 CONJECTURE"). *Suppose X is a closed, oriented, simply connected, smooth 4-manifold with even indefinite intersection form. Then*

$$\frac{b_2(X)}{|\operatorname{Sign}(X)|} \geq \frac{11}{8}.$$

Here $b_2(X)$ is the second Betti number, and $\operatorname{Sign}(X)$ the signature. The bound is realized by any connected sum of $K3$ surfaces.

CONJECTURE 9. *Every closed, oriented, smooth, simply connected 4-manifold is a connected sum of algebraic surfaces and their conjugates.*

(The conjugate surface has the reverse orientation. Also, S^4 is the empty connected sum.)

Conjecture 9 contains by implication the most famous unsolved problem in 4-manifold topology, the smooth Poincaré conjecture. This asserts that S^4 has only one smooth structure. Note that Donaldson's techniques, based on second homology, do not apply to topological 4-spheres.

The self-dual equations have been studied on certain noncompact manifolds. Building on this analytic foundation the Donaldson invariants can be defined for 4-manifolds X whose boundary is a *homology 3-sphere Y*.[1] We will discuss these ideas in broad terms here.[2] Let Y be a homology 3-sphere. Then there is a finite number of representations $\pi_1(Y) \to SU(2)$, all of which are irreducible except for the trivial representation (which requires special treatment in the theory). These representations correspond to flat connections on the trivial $SU(2)$ bundle over Y. Let C be the free abelian group generated by this finite set. Using spectral flow Floer [**F**] endows C with a $\mathbb{Z}/8$-grading. This uses the linearization of the self-dual equation. The full nonlinear equation enters in the definition of a differential $\partial : C_* \to C_{*-1}$, which renders C_* a chain complex. Its homology is the *Floer homology $HF_*(Y)$*. If $-Y$ denotes Y with the orientation reversed, then there is a pairing

$$(10) \qquad\qquad HF_*(Y) \otimes HF_*(-Y) \to \mathbb{Z}.$$

---

[1] Recall that a homology 3-sphere is a closed oriented 3-manifold with $H_1(Y;\mathbb{Z}) = 0$. A modified form of this discussion should extend to arbitrary closed oriented 3-manifolds Y. (For example, there is recent work for *rational* homology spheres.)

[2] For an introduction to Floer theory, see [**B**]. For more hard analysis, see [**T2**] and [**M**]. Atiyah's papers are always worth looking at [**A1**].

Now we describe the Donaldson invariants for manifolds bounded by a homology sphere. Let X be a compact oriented 4-manifold with boundary $\partial X = Y$ a homology 3-sphere. We study finite energy k-instantons on $X - \partial X$, choosing a metric so that ∂X is at infinite (as in Chapter 9). Finite energy requires the curvature to vanish asymptotically along Y (at infinity). This gives a flat connection on Y, or equivalently a representation of $\pi_1(Y)$. Thus for each representation $\alpha : \pi_1(Y) \to SU(2)$ we consider the moduli space $\mathcal{M}_k(\alpha)$ of self-dual connections which are asymptotic to the flat connection α on Y. The Donaldson invariant is then

$$(11) \qquad\qquad D_k(X) = \sum_\alpha [\mathcal{M}_k(\alpha)] \cdot \alpha.$$

Again this is our shorthand for the polynomials obtained by evaluating products of classes in the image of μ on $[\mathcal{M}_k(\alpha)]$. Hence $D_k(X)$ should be interpreted as a polynomial on $H_2(X)$ with values in $HF_*(Y)$. (Recall that α defines a class in $HF_*(Y)$.) Now suppose that we have a *closed* 4-manifold X smoothly cut along a homology sphere Y:

$$(12) \qquad\qquad X = X_1 \cup_Y X_2.$$

If $Y = S^3$ this is a smooth connected sum. Since $D_k(X_1)$ takes values in $HF_*(Y)$ and $D_k(X_2)$ takes values in $HF_*(-Y)$, these invariants can be paired using (10).

THEOREM 13 (DONALDSON). *If $b_2^+(X_i) > 0$ then $C_k(X) = \langle D_k(X_1), D_k(X_2)\rangle$.*

This result already contains Theorem 4, since $HF_*(S^3) = 0$. We caution the reader that these results are only in embryonic form. In particular, the reducible connections have not yet received a definitive treatment.

There is a heuristic interpretation of Floer homology which illuminates (10), (11), and Theorem 13. Namely, one can think of representing classes in $HF_*(Y)$ by oriented "middle dimensional cycles" in the space $\mathcal{C}_Y$ of all $SU(2)$ connections on Y modulo equivalence. A cycle representing a class in $HF_*(-Y)$ is roughly complementary to a cycle representing a class in $HF_*(Y)$, in the sense that the intersection is finite dimensional. Then the pairing (10) is the oriented intersection. If now X is a 4-manifold with boundary Y, the Donaldson invariant $D_k(X)$ is the "middle dimensional cycle" in $\mathcal{C}_Y$ consisting of boundary values of *all* self-dual connections on

X. In the situation (12), $D_k(X_1)$ and $D_k(X_2)$ are roughly complementary cycles in $\mathcal{C}_Y$, and the finite dimensional intersection is precisely the moduli space of self-dual connections on X. This is the content of Theorem 13.

Witten [**W1**] has argued that this latter description is reminiscent of a $3 + 1$ dimensional Euclidean Quantum Field Theory (QFT). In QFT one attaches to each closed oriented 3-manifold Y a vector space H_Y and to each oriented 4-manifold X an element $v_X \in H_{\partial X}$.[3] These elements are required to obey a property generalizing Theorem 13. In our situation there are some novelties. The vector spaces H_Y (the Floer homology) are finite dimensional, whereas in standard QFT they are infinite dimensional Hilbert spaces. Furthermore, these spaces and the elements v_X (the Donaldson invariants) are topological invariants, whereas in standard QFT they depend on continuous data. This is a manifestation of the fact that the Hamiltonian vanishes in these types of theory. Thus Floer homology and Donaldson invariants define a *topological* quantum field theory (TQFT) [**A2**]. Witten constructs this TQFT from a physical point of view. In related work he also recovers Gromov invariants of symplectic manifolds [**W2**] and Jones invariants of links in 3-space [**W3**]. Clearly this is only the beginning of a fruitful line of inquiry which we consider very exciting!

We have come full circle. In the late 1970's the *classical* Yang–Mills equations inspired deep mathematical studies, culminating in the work described here. These mathematical developments have in turn been related to a modified restrictive *quantum* theory, and may even shed light there.[4] We confidently replace the dire dead end we envisaged five years ago with an optimistic prediction that this field—in close collaboration with the physics—will continue to advance.

---

[3] If $\partial X = \emptyset$ then $H_{\partial X}$ is the field of scalars, usually the complex numbers.
[4] Especially in 2 dimensional conformal field theory [**W3**].

REFERENCES

[A1] M. F. Atiyah, *New invariants of 3 and 4 dimensional manifolds*, in "The Mathematical Heritage of Hermann Weyl," AMS Proceedings of Symposia in Pure Mathematics **48**, ed. R. O. Wells (1988), 285–299.

[A2] _______, *Topological Quantum Field Theories*, preprint, in honor of Rene Thom on his 65th birthday.

[B] P. Braam, *Floer homology groups for homology spheres*, Adv. Math. (to appear).

[D1] S. K. Donaldson, *The orientation of Yang–Mills moduli spaces and 4-manifold topology*, J. Diff. Geo. **26** (1987), 397–428.

[D2] _______, *Connections, cohomology and the intersection forms of 4-manifolds*, J. Diff. Geo. **24** (1986), 275–341.

[D3] _______, *Polynomial invariants for smooth four-manifolds*, preprint.

[D4] _______, *Irrationality and the h-cobordism conjecture*, J. Diff. Geo. **26** (1987), 141–168.

[D5] _______, *Infinite determinants, stable bundles, and curvature*, Duke Math. J. **54** (1987), 231–247.

[F] A. Floer, *An instanton invariant for three manifolds*, Commun. Math. Phys. **118** (1988), 215–40.

[FM1] R. Friedman and J. W. Morgan, *On the diffeomorphism types of certain algebraic surfaces I*, J. Diff. Geo. **27** (1988), 297–369; *II*, J. Diff. Geo. **27** (1988), 371–398.

[FM2] _______, *Algebraic surfaces and 4-manifolds: some conjectures and speculations*, Amer. Math. Soc. Bull. **18** (1988), 1–19.

[G] R. E. Gompf, *An infinite set of exotic $\mathbf{R}^4$'s*, J. Diff. Geo. **21** (1985), 283–300.

[M] G. Matic, *$SO(3)$-connections and rational homology cobordisms*, J. Diff. Geo. **28** (1988), 277–307.

[Si] C. Simpson, *Constructing variations of Hodge structure using Yang–Mills theory and applications to uniformization*, Journal of the American Mathematical Society **1** (1988), 867–918.

[T1] C. H. Taubes, *Gauge theory on asymptotically periodic 4-manifolds*, J. Diff. Geo. **25** (1987), 363–430.

[**T2**] __________, *Casson's Invariant and Gauge Theory*, to appear in J. Diff. Geo.

[**UY**] K. Uhlenbeck and S. T. Yau, *On the existence of Hermitian-Yang-Mills connections in stable vector bundles*, Comm. Pure Appl. Math. **XXXIX** (1986), S257–93.

[**W1**] E. Witten, *Topological quantum field theory*, Commun. Math. Phys. **117** (1988), 353–386.

[**W2**] ________, *Topological Sigma Models*, Commun. Math. Phys. **118** (1988), 411–449.

[**W3**] ________, *Quantum field theory and the Jones polynomial*, Commun. Math. Phys. **121** (1989), 351–99.

CONTENTS

INTRODUCTION TO THE FIRST EDITION

Topologists study three types of manifolds—topological or continuous (TOP), piecewise linear (PL), differentiable ($DIFF$)—and the relationships among them. A basic problem is to ascertain when a topological manifold admits a PL structure and, if it does, whether there is also a compatible smooth structure. By the early 1950's it was known that every topological manifold of dimension less than or equal to three admits a unique smooth structure. In 1968 Kirby and Siebenmann determined that for a topological manifold M of dimension at least five, there is a single obstruction $\alpha(M) \in H^4(M; \mathbb{Z}_2)$ to the existence of a PL structure. There are further discrete obstructions to lifting from PL to $DIFF$; these have coefficients in groups of homotopy spheres. Fortunately, a simplification in dimension four absolves us from having to consider the piecewise linear category again: Every PL 4-manifold carries a unique compatible differentiable structure. Now the Kirby–Siebenmann obstruction $\alpha(M)$, which lives on the 4-skeleton of an n-manifold M, relates in special cases to a result of Rohlin dating back to 1952. Rohlin's Theorem states that the signature of a smooth spin 4-manifold is divisible by 16. The arithmetic of quadratic forms shows that the signature of a topological "spin" ($=$ almost parallelizable) 4-manifold M is divisible by 8, and $\alpha(M) \in \mathbb{Z}_2 = 8\mathbb{Z}/16\mathbb{Z}$ is the signature $\mod 16$. If M is not spinable, the Kirby–Siebenmann invariant is an extra piece of information not related to the intersection form.

Recently, a new type of "obstruction" to the smoothability of 4-manifolds was discovered by Simon Donaldson. He proved that if the intersection form ω of a compact, simply connected *smooth* 4-manifold is definite, then ω is equivalent over the integers to the standard diagonal form $\pm\mathrm{diag}(1, 1, \ldots, 1)$. One year earlier Michael Freedman had classified all compact, simply connected *topological* 4-manifolds, and he found that every unimodular symmetric bilinear form is realized as the intersection form of some topological 4-manifold. Together these results give many examples of nonsmoothable 4-manifolds with vanishing Kirby–Siebenmann invariant. Freedman and others saw that Donaldson's Theorem, in view of work done by Andrew Casson and others in the early 1970's, implies an even more striking result: the existence of exotic differentiable structures on $\mathbb{R}^4$. At this time it is not known how many such fake $\mathbf{R}^4$'s exist, although several have been found. According to Freedman, topologists speculate that there may be an uncountable number. If this turns out to be true, then the classification of smooth structures, which in higher dimensions is accomplished with characteristic classes and is therefore a discrete problem, could stray into the realm of geometry; just as there are (continuous) moduli spaces of complex structures on Riemann surfaces, so too there may be the moduli spaces of smooth structures on 4-manifolds! Regardless, Donaldson's Theorem makes clear the impossibility of characterizing smooth structures in four dimensions in terms of bundle lifting (i.e., characteristic classes). As concrete examples where bundle lifting fails, we cite $|E_8 \oplus E_8|$ and fake $\mathbf{R}^4$. It is striking that Rohlin's Theorem and Donaldson's Theorem can both be proved by studying a class of decidedly nondiscrete objects: elliptic operators on smooth 4-manifolds. In fact, it remains a challenge for topologists to find a proof of Donaldson's Theorem which does not rely so heavily on geometry and analysis.

The study of elliptic operators on compact manifolds often leads to theorems relating the geometry of the manifold to its topology. We begin with the cornerstone of linear elliptic theory, the Hodge–de Rham Theorem. A smooth n-manifold M comes equipped with a natural elliptic complex of differential operators

$$O \longrightarrow \Omega^0(M) \xrightarrow{d} \Omega^1(M) \xrightarrow{d} \ldots \xrightarrow{d} \Omega^n(M) \longrightarrow 0,$$

where at the qth stage $d : \Omega^q(M) \to \Omega^{q+1}(M)$ is exterior differentiation from q-forms to $(q+1)$-forms. For compact M this de Rham complex has finite dimensional cohomology groups

$$(1) \qquad H^q_{DR}(M) = \frac{\mathrm{Ker}\; d : \Omega^q(M) \to \Omega^{q+1}(M)}{\mathrm{Im}\; d : \Omega^{q-1}(M) \to \Omega^q(M)}$$

which are isomorphic to the real singular cohomology groups $H^q(M; \mathbf{R})$. Hence these spaces $H^q_{DR}(M)$, which a priori depend on the differentiable structure, are actually invariants of the topological structure. When M has a Riemannian metric, there is a canonical representative of each cohomology class. This is chosen by minimizing the energy

$$\mathcal{E}(\alpha) = \int_M |\alpha|^2$$

over a given cohomology class (i.e., over $\alpha = \alpha_\circ + d\beta$ where $\beta \in \Omega^{q-1}(M)$ and $\alpha_\circ$ is any closed q-form in the given class). The Hodge–de Rham Theorem states that in each cohomology class there is a unique minimizing α, which satisfies the Euler–Lagrange equations

$$(2) \qquad\qquad d^*\alpha = 0.$$

Since we also have $d\alpha = 0$, equation (2) is equivalent to

$$(3) \qquad\qquad \Delta\alpha = (dd^* + d^*d)\alpha = 0.$$

Here $\Delta = dd^* + d^*d$ is the Laplace–Beltrami operator on forms. Any α satisfying (3) is called harmonic. Applications of Hodge–de Rham Theory to global differential geometry often obtain by expressing the difference of the Laplace operator on forms, $dd^* + d^*d$, and a differential operator formed from the full covariant derivative, $\nabla^*\nabla$, as an algebraic operator involving curvature. Applying this to 1-forms, for example, Bochner proved in 1946 that $H^1(M;\mathbf{R}) = 0$ for compact M which carry a metric of positive Ricci curvature.

Hodge–de Rham Theory extends to more general linear elliptic operators. An elliptic complex is a finite sequence of (first order) operators

$$0 \longrightarrow C^\infty(\xi_\circ) \xrightarrow{D_1} C^\infty(\xi_1) \xrightarrow{D_2} \ldots \xrightarrow{D_r} C^\infty(\xi_r) \longrightarrow 0$$

between vector bundles ξ_i over M such that (i) $D_{i+1} \circ D_i = 0$, and (ii) on the symbol level

$$0 \longrightarrow \xi_0 \xrightarrow{\sigma(D_1)(\theta)} \xi_1 \xrightarrow{\sigma(D_2)(\theta)} \ldots \xrightarrow{\sigma(D_r)(\theta)} \xi_r \longrightarrow 0$$

is exact for nonzero $\theta \in T^*M$. Generalized cohomology groups $H^q(\xi)$ are defined as in (1), and for compact M these are finite dimensional. If metrics on ξ_i and a volume form on M are given, the (formal) L^2 adjoints D_q^* are defined. The generalized Hodge–de Rham Theorem says that again there is a unique canonical representative f in each cohomology class satisfying

$$D_{q-1}^* f = 0,$$

or equivalently, since $D_q f = 0$ also,

$$(D_{q-1} D_{q-1}^* + D_q^* D_q)f = 0.$$

The Atiyah–Singer Index Theorem expresses the alternating sum $\sum_q (-1)^q \dim H^q(\xi)$ in terms of characteristic classes of M and ξ constructed from the symbol sequence. (To determine a particular $\dim H^q(\xi)$, one usually combines this with vanishing theorems.)

Elliptic complexes can be used to explore the relationship between differential geometry, algebraic geometry, and topology. Of immediate interest is a particular application involving only topology: Rohlin's Theorem. On a spin manifold M there is a natural elliptic operator, the Dirac operator, whose index is the $\hat{A}$-genus of M. This is a certain characteristic class of M evaluated on the fundamental cycle, and for a 4-manifold it turns out to be $\frac{1}{8}$ times the signature. Since the index of an elliptic complex is an integer, the $\hat{A}$-genus of M is integral. (It was precisely this problem—to explain the integrality of $\hat{A}(M)$ for spin manifolds M—which led Atiyah and Singer to the Index Theorem.) Furthermore, the spin representation in four dimensions is symplectic, and thus the space of harmonic spinors (the kernel and cokernel of the Dirac operator) is quaternionic. It follows that $\hat{A}(M)$ is an *even* integer, and the signature of M is divisible by 16.

In four dimensions there is an important twisted Dirac operator obtained by tensoring with one of the half-spin bundles. We mention it here as it is essentially a linearized version of the nonlinear operator Donaldson studies to deduce his topological result. This operator can be described explicitly in terms of self-duality and differential forms. Namely, if M is an oriented Riemannian 4-manifold, then the six dimensional bundle $\Lambda^2 M$ splits canonically into the sum of three dimensional bundles $\Lambda^2 M = \Lambda^2_+ M \oplus \Lambda^2_- M$. This corresponds to the Lie algebra decomposition $\mathfrak{so}(4) = \mathfrak{so}(3) \oplus \mathfrak{so}(3)$. We get a new elliptic complex

$$(4) \qquad 0 \longrightarrow \Omega^0(M) \xrightarrow{d} \Omega^1(M) \xrightarrow{d_-} \Omega^2_-(M) \longrightarrow 0$$

by composing $d : \Omega^1(M) \to \Omega^2(M)$ with the projection $P_- : \Omega^2(M) \to \Omega^2_-(M)$. Then the twisted Dirac operator is $d^* \oplus \sqrt{2} d_- : \Omega^1(M) \to \Omega^0(M) \oplus \Omega^2_-(M)$.

Nonlinear analysis has had as great an impact on geometry and topology as linear analysis. Basic results come from the Morse Theory of geodesics on Riemannian manifolds, a variational theory for nonlinear ordinary differential equations. One of the first applications is the Hadamard–Cartan Theorem (1898/1928) which asserts that the universal cover of a complete Riemannian n-manifold of nonpositive curvature is diffeomorphic to $\mathbf{R}^n$.

For positively curved manifolds we have Sumner Byron Myer's Theorem (1941): A complete Riemannian manifold with positive Ricci curvature is compact and has finite fundamental group. This is a stronger result than is obtained from linear theory, since Bochner's Theorem assumes M compact and only concludes $H^1(M;\mathbf{R}) = 0$. More spectacular is the use of Morse Theory by Bott in 1956 to study geodesics on Lie groups, which led him to his celebrated Periodicity Theorem.

Important applications of nonlinear elliptic *partial* differential equations to geometry and topology lagged behind until very recently. In the last five years a number of results have appeared, many using techniques involving minimal surfaces. For example, Schoen and Yau's proof of the positive mass conjecture in general relativity, which relies on properties of the minimal surface equation, yields the following geometric by-product: if the fundamental group of a 3-manifold M contains a subgroup isomorphic to the fundamental group of a compact surface with genus ≥ 1, then M admits no metric of positive scalar curvature. At about the same time Meeks and Yau used minimal surfaces to give new proofs of Dehn's Lemma (the Loop Theorem) and the Sphere Theorem, two fundamental results in 3-manifold topology. More importantly, they proved a new theorem—the Equivariant Loop Theorem—which, added to work of Thurston, Bass, and others, completed a proof of the Smith Conjecture, a longstanding open problem about $\mathbf{Z}_n$ actions on S^3. Recent work of Freedman and Yau examine more general group actions on S^3 using minimal surface techniques. Alan Edmunds has recently given a purely topological proof of the Equivariant Loop Theorem. However, for a theorem of Meeks, Simon, and Yau of the same vintage—if a 3-manifold has no fake cell (counterexample to the Poincaré conjecture), then its universal cover has no fake cell—there is still no purely topological proof. Of all applications of analysis to topology via geometry, the Equivariant Loop Theorem and its consequences in 3-manifold topology bear the closest relationship to Donaldson's Theorem in 4-manifold topology. The same low dimensional topologists who were learning about minimal surfaces in 3-manifolds a few years ago are now studying the Yang–Mills equations on 4-manifolds.

Even with hindsight afforded by the passage of time, it is difficult to find a pattern in the important applications of analysis to topology, and to make predictions for the future would be foolhardy. Nevertheless, our brief historical survey omitted applications of partial differential equations to the geometry and topology of complex manifolds, which are even more numerous than applications to differentiable manifolds. In fact, an extension of the self-dual equations Donaldson uses can be used to study stable holomorphic vector bundles over complex Kähler manifolds.

We can formulate Yang–Mills as a nonlinear generalization of Hodge Theory. In addition to a Riemannian 4-manifold M, we also start with a normed vector bundle η. We set up a variational problem for connections D on η by taking as action the energy (L^2 norm) of the curvature F_D:

$$(5) \qquad \mathcal{YM}(D) = \int_M |F_D|^2.$$

A critical point of this Yang–Mills functional satisfies the Euler–Lagrange equations

$$(6) \qquad D^* F_D = 0,$$

a nonlinear generalization of (2). (Recall that curvature is a quadratic expression in the connection, so the nonlinearity is mild.) In view of the Bianchi identity $DF_D = 0$, we also get a Laplace-like equation

$$(DD^* + D^*D)F_D = 0.$$

The second order Yang–Mills equations (6) are automatically satisfied by solutions to first order equations which yields absolute minima of (5). These are the self-dual (anti-self-dual) equations

$$(7) \qquad *F_D = \pm F_D.$$

Donaldson's Theorem, stated above, gives a restriction on the topology of a compact, simply connected smooth four-manifold M. The theorem is proved by studying the solutions of the nonlinear semi-elliptic system of equations (7). The operators involved in the equations are nonlinear generalizations of the four-manifold Dirac operator described earlier, and as such are special to four dimensions. The space of solutions is divided out by a natural equivalence to produce the "moduli space" $\mathcal{M}$. As with linear elliptic systems, we learn about the topology of M by studying the geometry of the solution space, only now that study is much more involved— in the linear case the solutions form a vector space, and the geometry is completely determined by its dimension. For the self-dual equations, roughly speaking, the moduli space $\mathcal{M}$ is an oriented five-manifold with point singularities, neighborhoods of the singular points are cones on $\mathbf{CP}^2$, and M appears as the boundary of $\mathcal{M}$. Now the argument proceeds using cobordism. Remarkable is how neatly each bit of topological information on M fits the analysis! The positivity of the intersection form is necessary for Taubes' existence theorem. Our proof that $\mathcal{M}$ is orientable and the fact that $\dim \mathcal{M} = 5$ both require that the first Betti number of M vanish. The ends of $\mathcal{M}$ can be identified as $\mathbf{R} \times M$, and postulating $\pi_1(M) = 0$ ensures that there is only one end. The proof works for exactly the hypotheses given, and basically for no other.

Due to this fine tuning between the analysis and topology, the directions in which Donaldson's Theorem can be extended are very limited, although there are possibilities open for treating 4-manifolds with singularities or with boundary. Nevertheless, all the evidence indicates that gauge theory is here to stay, both in mathematics and in physics. There are several quite different reasons why gauge theory is important in mathematics, aside from the application discussed here. One is the beautiful dichotomy between the algebraic twistor description of self-dual fields over self-dual 4-manifolds and the nonlinear analysis. Here $\mathcal{M}$ can be studied with tools from algebraic geometry, quaternionic linear algebra, and nonlinear PDE. In a similar vein, holomorphic bundles over complex Kähler manifolds of all dimensions can be examined using an extension of the self-dual equations. Atiyah and Bott have already investigated the topology of the moduli space of stable vector bundles over Riemann surfaces in this framework. The three dimensional Yang–Mills equations remain a challenge. Although abstract existence theorems guarantee solutions, their geometrical significance has yet to be determined. Finally, the equations themselves, particularly when coupled with an external "matter field" (the Yang–Mills–Higgs equations), are really interesting PDE's. Not only is there motivation from physics to study them, but their topological and geometric features are both conceptually and technically fascinating.

Because our exposition draws on three branches of mathematics—topology, geometry, and analysis—we have endeavored to supply background material whenever possible. The following chapter by chapter description will enable the reader to make his own roadmap through the book.

In §1 we discuss both topological and differentiable four-manifolds. Three equivalent definitions of the intersection form are given. At the end of this chapter we sketch Freedman's argument for the existence of a fake $\mathbf{R}^4$.

The basic geometry of gauge theory is set up in §2. We choose to work with vector bundles rather than principal bundles in order that concrete formulas be expressed. Perhaps some geometric insight into connections is lost, though, and we take this opportunity to explain the covariant derivative with pictures. Consider the simplest case of real-valued functions f on $\mathbf{R}^2$. A basic principle of modern differential geometry is simply this: We understand functions (or sections of bundles) by studying the geometry of their graphs.

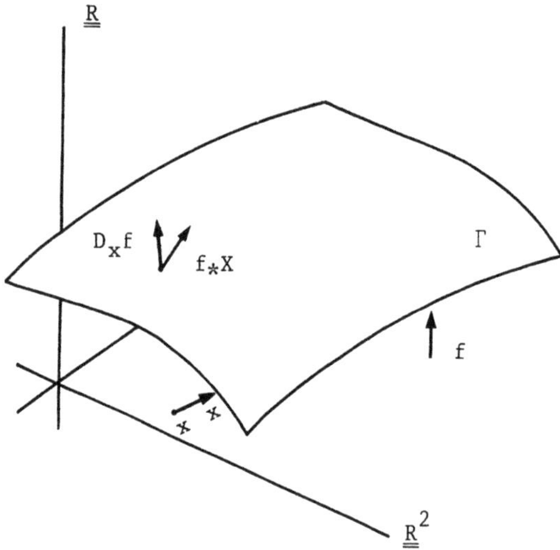

In this spirit the directional derivative $D_X f$ of f in the direction X can be computed by first lifting X to a tangent vector $f_* X$ to $\Gamma = \operatorname{graph} f$. Then the vertical part of $f_* X$ measures the rate of change of f in the direction X. By identifying the tangent space to $\mathbf{R}$ with $\mathbf{R}$, we have determined $D_X f$. In this example the vertical projection, fixed by specifying its kernel, the horizontal subspace at $f(x)$, is given canonically by the product structure of $\mathbf{R}^3 = \mathbf{R}^2 \times \mathbf{R}$. Over topologically nontrivial manifolds there are vector bundles which are not products, and then the horizontal distribution, or connection, must be chosen as an additional piece of geometric data. The obstruction to a local basis of flat sections is the curvature of the connection, and global properties of the curvature reflect the twisting of the bundle.

We study connections satisfying a particular system of differential equations. The set of all connections on a bundle forms an affine space $\mathcal{A}$ (the difference of two connections is a tensor field on the base), and the group $\mathcal{G}$ of bundle automorphisms acts naturally on $\mathcal{A}$. The Yang–Mills equations are invariant under this action. Therefore, our moduli space $\mathcal{M}$ is taken to be a subset of $\mathcal{A}/\mathcal{G}$, where it is finite dimensional. At the end of §2 we prove Donaldson's Theorem modulo the topological properties of $\mathcal{M}$ demonstrated in later chapters.

For a generic metric on M, the moduli space is a smooth 5-manifold with a finite number of singular points. Our approach in §3 and §4 differs from Donaldson's. His perturbation of $\mathcal{M}$ is not induced by a perturbation of the metric, and his more abstract setup leads to a somewhat simpler argument. On the other hand, the space we end up with is still the space of solutions to the Yang–Mills equations, but now the base metric is perturbed. Both proofs use the Sard–Smale Theorem to construct perturbations. We treat irreducible connections in §3. The singular points of $\mathcal{M}$ correspond to reducible connections, and near these points $\mathcal{M}$ looks like a cone on $\mathbf{CP}^2$. In §4 we redo the genericity theorem taking into account the extra symmetry provided by the S^1 holonomy of a reducible connection.

The arguments of §5 are mostly topological. The index bundle of our nonlinear version of (4) is an extension of the tangent bundle $T\mathcal{M}$, and its existence allows us to deduce the orientability of $\mathcal{M}$ from the simple connectivity of $\mathcal{A}/\mathcal{G}$. This, in turn, follows from the connectedness of $\mathcal{G}$. The path group of $\mathcal{G}$ turns out to be the set of homotopy classes $[M, S^3]$, and this can be computed from the Steenrod Classification Theorem. A more geometric argument based on Pontrjagin's Construction is given in Appendix B.

§6 is an odd mix of analysis and geometry. Only the grafting procedure is part of Taubes' Theorem; the rest is background material. We begin with a geometric description of the moduli space of instantons on S^4. Because the conformal group preserves the Yang–Mills equations and acts transitively on S^4, our presentation emphasizes its role. Instantons on S^4 can be localized by homothety, and Taubes' ingenious idea is to transfer these to M. After we describe this grafting procedure, we turn to tools from analysis that will be used to complete the proof.

In §7 analysis is at the fore. Novel is blowing up the metric to compensate for a singularity in the curvature. (This technique has also been used to handle singularities in other PDE problems.) Nevertheless, nothing in this world is free, and in this case we pay by being forced to work on a noncompact manifold. Our control over the blow-up process, and hence the noncompactness, is exhibited in our estimates, which enable us to complete the proof of Taubes' Theorem. We also easily obtain a local connectivity result needed later. By working on our blown-up, noncompact manifold, we obtain an argument much simpler than Donaldson's. Indeed, this was the motivation for giving the different proof of Taubes' Theorem.

The compactness of the moduli space is proved in §8. Solutions to the self-dual equations are regular, and as our argument applies to nonlinear elliptic equations in general, we give all of the details in the Regularity Theorem. To carry out the proof we exploit a canonical local splitting of the $\mathcal{G}$ action on $\mathcal{A}$ obtained from PDE, the so-called Coulomb gauge. The gauge fixing lemma, as well as the patching argument needed to complete the proof of the Compactness Theorem, are omitted since we could not improve the published proofs. In §8 we also include a long, technical prescription for measuring the concentrated curvature of localized instantons, which is crucial in §9. In fact, the map $\overline{B}$ specifying the center and scale is the collar of M in the moduli space. Some estimates on concentrated instantons appear as dividends of our compactness arguments.

In §9 we follow Donaldson's original proof of the Collar Theorem. Here again our blown-up version of M makes several arguments easier. In the first section we continue to discuss the structure of concentrated instantons, now in an annular region near the center. The decay estimates derived here also lead to a quick proof of the Removable Singularities Theorem, which we provide in Appendix D. As always, we understand the collar by analogy with concentrated instantons on S^4; therefore, it is easily ascertained that the five dimensional tangent space at a concentrated instanton consists of infinitesimal almost conformal deformations. Our proof that $\overline{B}$ is a local diffeomorphism makes this precise. Combining with the compactness results of §8, we then conclude that $\overline{B}$ is a finite covering map. To prove that $\overline{B}$ is 1:1 we require exponential gauges, which we explain in detail. Finally, we patch together our various estimates and invoke the local connectivity result of §7 to complete the proof of the Collar Theorem.

There is a new proof of special cases of Donaldson's Theorem due to the topologists Fintushel and Stern. Their methods do not apply to all intersection pairings, but do apply to manifolds with nontrivial fundamental group. The new insight is that by replacing the $SU(2)$ bundle Donaldson uses with an appropriate $SO(3)$ bundle, the resulting moduli space of instantons is one dimensional and compact. Now a simple count of boundary points replaces the cobordism argument. We discuss their technique in §10.

There are five appendices. The first provides some technical arguments involving Sobolev spaces that would have burdened the exposition in §3. A pleasing, geometric computation of $[M, S^3]$ is included in Appendix B. Appendix C is a discussion of Weitzenböck formulas from a general point of view emphasizing the role of the orthogonal group. Our setup is essentially an intrinsic way of calculating with geodesic normal coordinates. As an antidote to this abstraction, we derive a particular Weitzenböck formula with moving frames. The exact coefficients in this formula are crucial for the decay estimates in §9. As already mentioned, Appendix D is a proof of the Removable Singularities Theorem using these estimates. We include various topological arguments, including the classification of $U(1)$, $SU(2)$, and $SO(3)$ bundles, in Appendix E.

Dirac Operator   This is the most fundamental differential operator on a spin manifold. In even dimensions Dirac is a first order operator $\not\partial$: $C^\infty(S^+) \to C^\infty(S^-)$ between the "half-spin" bundles, and in four dimensions the complex (4) is obtained by tensoring with S^- to obtain $\not\partial_{S^-} : C^\infty(S^+ \otimes S^-) \to C^\infty(S^- \otimes S^-)$. See [**AHS**] for details.

Elliptic Theory   There is a very accessible exposition of the de Rham Theorem in [**Wa**]. Palais' book [**P2**] covers the basic theory of Fredholm operators, as well as the Atiyah–Singer Index Theorem.

End   Loosely speaking the ends of a manifold M are the parts which extend out to infinity.

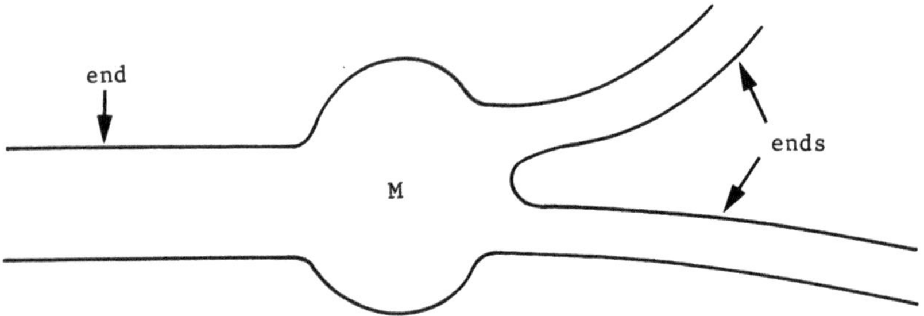

More formally, the collection $\{M \backslash K\}$ of complements of compact subsets $K \subseteq M$ is directly by inclusion, and the set of ends is precisely the inverse limit of this directed set [**St2**, p. 3].

Kirby–Siebenmann invariant and Smoothing Theory   Let $DIFF_n$ be the group of diffeomorphisms of $\mathbf{R}^n$, PL_n the group of invertible PL maps of $\mathbf{R}^n$, and TOP_n the group of homeomorphisms of $\mathbf{R}^n$. Then there are inclusions

$$DIFF_1 \subset DIFF_2 \subset \dots \subset DIFF_n \subset \dots$$
$$PL_1 \subset PL_2 \quad \subset \dots \subset PL_n \quad \subset \dots$$
$$TOP_1 \subset TOP_2 \subset \dots \subset TOP_n \subset \dots ,$$

and we can define limit spaces $DIFF$, PL, and TOP. (This construction is the same as for Lie groups. In fact, the stable orthogonal group O is homotopy equivalent to $DIFF$.) Furthermore, as for Lie groups, these topological groups have classifying space (cf. Appendix E), denoted $BDIFF$, BPL, and $BTOP$, and there are maps $BDIFF \to BPL \to BTOP$ corresponding to the (nontrivial) fact that a differentiable manifold is PL is topological. A topological manifold M carries a topological tangent bundle. It is represented by a map $\tau : M \to BTOP$, and one might hope that liftings of τ to BPL (respectively, $BDIFF$) correspond to PL (differentiable) structures on M. In fact, the main result of smoothing theory states that this is true in dimensions ≥ 5 (if $\partial M = \emptyset$; otherwise in dimensions ≥ 6). The precise theorem states that homotopy classes of lifts correspond to isotopy classes of $PL(DIFF)$ structures. Thus smoothing theory in these dimensions is reduced to obstruction theory. Now the fiber of $BPL \to BTOP$ is TOP/PL, and in 1969 Kirby and Siebenmann determined that the only nontrivial homotopy group of this fiber is $\pi_3(TOP/PL) = \mathbb{Z}_2$. The image under τ^* of the corresponding universal obstruction in $H^4(BTOP; \pi_3(TOP/PL))$ is the Kirby–Siebenmann invariant $\alpha(M) \in H^4(M; \mathbb{Z}_2)$. Obstructions to smoothing a PL manifold arise from the homotopy groups $\pi_m(PL/DIFF) = \Theta_m (m \geq 5)$, where Θ_m is the Kervaire–Milnor group of oriented isomorphism classes of smooth homotopy m-spheres. The first few values are

$$\Theta_m = 0, 0, \mathbb{Z}_{28}, \mathbb{Z}_2, \mathbb{Z}_2 \oplus \mathbb{Z}_2, \mathbb{Z}_6, \mathbb{Z}_{992} \quad \text{for} \quad m = 5, 6, \ldots, 11.$$

Smoothing Theory in dimension four is not completely known; Donaldson's Theorem is a big advance. However, Cerf [**Ce**] proved that every PL 4-manifold carries a unique compatible differentiable structure. In dimensions one, two, and three all topological manifolds are uniquely smoothable [**Mo**]. A general reference for smoothing theory, which includes an extensive bibliography, is [**KS**].

Morse Theory   Milnor's book [**M5**] is a classic.

PL Manifold   We first need the notion of a PL map. Let $U \subseteq \mathbf{R}^n$ and $V \subseteq \mathbf{R}^n$ be open sets, and suppose $f : U \to V$. Then f is a PL map if there exists a subdivision of the standard rectilinear triangulation (in both the domain and range) for which f is linear on each simplex. A topological manifold M has a *PL* structure if there is an atlas on M for which the transition functions are *PL* maps. (Compare with the definition of "differentiable manifold" in §1.)

Rohlin's Theorem   A topological proof along the lines of Rohlin's original idea is given in [**FK**].

Three-Manifold Topology and Minimal Surfaces   The three basic theorems of three-manifold topology—Dehn's Lemma, the Loop Theorem, and the Sphere Theorem—were proved in 1957 by Papakyriakopoulos. (Dehn stated his result in 1910, but gave a misproof.) Let M be a 3-manifold with boundary which, by Smoothing Theory, we can take to be smooth. Dehn's Lemma states that if $\gamma \subset \partial M$ is a Jordan curve which is contractible in M, then γ bounds an embedded disk in M. The Loop Theorem refers to the case where the embedded curve γ is an essential loop in ∂M. Combining with Dehn's Lemma, we have the usual formulation: If the kernel of $\pi_1(\partial M) \to \pi_1(M)$ is nontrivial, then there exists an embedded disk in M whose boundary lies in ∂M and represents a nontrivial element of this kernel. Finally, the Sphere Theorem states that if M is orientable and $\pi_2(M) \neq 0$, then there is an embedded S^2 in M which is not contractible. Meeks and Yau derived geometric versions of these theorems for compact manifolds. Namely, for suitable metrics on M (which always exist) they proved that the embedded disks in Dehn's Lemma and the Loop Theorem, as well as the embedded S^2 in the Sphere Theorem, can be chosen to be least area minimal embeddings. Furthermore, the images of any two such embeddings are either identical or intersect only along the boundary (are disjoint in the Sphere Theorem). There are equivariant formulations of these geometric theorems which lead to a proof of the Smith Conjecture: If $\mathbf{Z}_n$ acts on S^3 with a one dimensional fixed point set F, then F is an unknotted circle. The reader can consult [**He**], [**St2**] for three-manifold topology and [**MY1**], [**MY2**], [**MSY**] for the results involving minimal surfaces. We remark that in both contexts there are sharper versions of the three basic theorems which reflect the group theory of the manifold more closely.

Topological Spin Manifold   Simply a manifold whose first two Stiefel–Whitney classes vanish. (Stiefel–Whitney classes for topological manifolds are defined using Steenrod squares [**Sp**, §6.10], [**MS**, §8].)

14

§1 FAKE $\mathbf{R}^4$

In this chapter we give an account of the topological ideas leading to the existence of a "fake $\mathbf{R}^4$". What distinguishes $\mathbf{R}^4_{\text{fake}}$ is its differentiable structure. After first reviewing the notion of a differentiable structure on a manifold, we describe the algebraic invariants used to classify topological 4-manifolds. Not all 4-manifolds admit a smooth structure, and specific nonexistence results, including Donaldson's Theorem, are stated. Finally, all of this is tied together by a sketch of the proof that an exotic differentiable structure exists on $\mathbf{R}^4$. We refer the reader to [**Fr2**] for another expository account of this material.

Differentiable Structures.

Let M be a topological manifold of dimension n. Thus M is a topological space which looks locally like ordinary Euclidean space $\mathbf{R}^n$. On M we know what open sets and continuous functions are, but we can't do calculus. So we want to add information, compatible with the topology of M, which allows us to distinguish a subring of smooth functions in the ring of all continuous functions on M. To accomplish this we specify a smoothly compatible set of *coordinate charts*. A *differentiable structure* on M, then, is a covering of M by open sets $\{U_\alpha\}$ and homeomorphisms $\phi_\alpha : U_\alpha \to \mathbf{R}^n$ so that the *transition functions*

$$\phi_\beta \circ \phi_\alpha^{-1} : \phi_\alpha(U_\alpha \cap U_\beta) \longrightarrow \phi_\beta(U_\alpha \cap U_\beta)$$

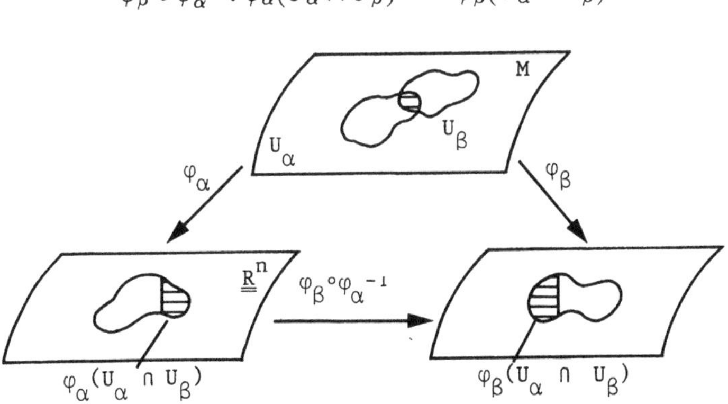

are smooth. This makes sense as $\phi_\alpha(U_\alpha \cap U_\beta)$ and $\phi_\beta(U_\alpha \cap U_\beta)$ are open subsets of $\mathbf{R}^n$, and we know how to differentiate functions between such sets by taking limits of difference quotients. Of course, we now say that $f : M \to \mathbf{R}$ is smooth if $f \circ \phi_\alpha^{-1} : \phi_\alpha(U_\alpha) \to \mathbf{R}$ is smooth for each U_α. This defines a ring of smooth functions on M.

As a first example consider $M = \mathbf{R}^n$. We reiterate that here differentiation already makes sense: compute difference quotients! In terms of our definition we cover M by one open set $(U = M)$ and take ϕ to be the identity map. Denote $\mathbf{R}^n$ together with this differentiable structure by $\mathbf{R}^n_{\mathrm{std}}$. It is not difficult to construct different differentiable structures. For example, take $N = \mathbf{R}$, cover N by one open set as before, but now define $\phi : N \to \mathbf{R}$ by $\phi(x) = x^{1/3}$. Notice that ϕ is a homeomorphism. So ϕ defines a new notion of smoothness on N—a function $f : N \to \mathbf{R}$ is smooth in the new sense if and only if $f \circ \phi^{-1} : \mathbf{R} \to \mathbf{R}$ is smooth in the sense of difference quotients. For example, the function $f(x) = x^{2/3}$ is smooth on N since $f \circ \phi^{-1}(y) = y^2$ is smooth in the ordinary sense. We have succeeded, then, in defining a different notion of smoothness on $\mathbf{R}$. But our new notion is not very different. In fact, the map ϕ (regarded as a map $N \to \mathbf{R}^n_{\mathrm{std}}$) is a *diffeomorphism* between N and $\mathbf{R}_{\mathrm{std}}$. Now it is conceivable that a more complicated covering of $\mathbf{R}$ by many coordinate charts could produce a truly different differentiable structure. This is not possible, though. For if X denotes $\mathbf{R}$ with some given differentiable structure, then by a partition of unity we can construct a complete Riemannian metric on X. Now it is easy to see that the exponential map $\exp : T_0 X \to X$ is a diffeomorphism. But $T_0 X$, the tangent space to X at the origin, is $\mathbf{R}_{\mathrm{std}}$. So X is diffeomorphic to $\mathbf{R}_{\mathrm{std}}$.

Thus on the real line there are no exotic notions of differentiability. The situation is different for more complicated spaces. In 1956 John Milnor provided a concrete example by constructing an exotic differentiable structure on S^7 [**M1**]. Since then exotic structures on higher dimensional spheres have also been discovered, and the classification of these structures has been reduced to problems in homotopy theory [**KM2**]. As one might expect, the flat spaces $\mathbf{R}^n$ are tamer. Indeed, it is possible to show that for $n \neq 4$ any differentiable structure on $\mathbf{R}^n$ is equivalent to the standard one [**Mo**], [**St**]. But there is the following

AMAZING FACT. *There is an exotic differentiable structure on* $\mathbf{R}^4$.

The existence of this $\mathbf{R}^4_{\mathrm{fake}}$ is proved by a remarkable combination of topology, analysis, and geometry. We will deal with the analysis and geometry later; in this chapter we discuss some topological aspects.

For now we are content to describe a striking property of $\mathbf{R}^4_{\text{fake}}$ which forbids a diffeomorphism to $\mathbf{R}^4_{\text{std}}$:

(1.1)   *There exists a compact set C so that no smoothly embedded 3-sphere in $\mathbf{R}^4_{\text{fake}}$ surrounds C.*

Of course, in $\mathbf{R}^4_{\text{std}}$ there are arbitrarily large smooth 3-spheres. Also, since $\mathbf{R}^4_{\text{fake}}$ is *homeomorphic* to ordinary $\mathbf{R}^4$, there are arbitrarily large *continuously* embedded 3-spheres in $\mathbf{R}^4_{\text{fake}}$. (Just choose any norm and consider $\{x \in \mathbf{R}^4_{\text{fake}} : \|x\| = R\}$ for large R.) So in the differentiable structure of $\mathbf{R}^4_{\text{fake}}$, spheres near ∞ are very jagged.

Topological 4-Manifolds.

Throughout this section M denotes a compact, simply connected four dimensional topological manifold. In particular, M is orientable and we fix a definite orientation. Construct a symmetric bilinear form on $H_2(M; \mathbf{Z})$ as follows. Suppose $a, b \in H_2(M; \mathbf{Z})$ can be represented by oriented surfaces A, B in M which intersect transversely. (This is always possible if M is smooth—see $(E.9)$ in Appendix E. For topological manifolds we are forced to adopt the definition in the next paragraph.) Define

$$\omega(a, b) = A \cdot B$$

where $A \cdot B$, is the (oriented) intersection number of the cycles A and B.

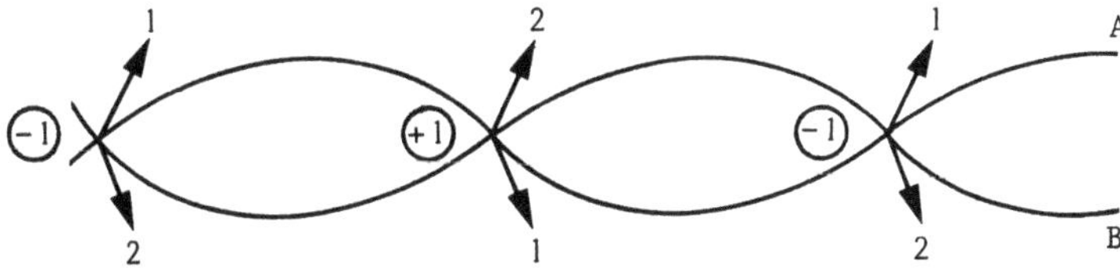

The form ω is symmetric as both A and B are 2 dimensional. Also, ω is *unimodular* (that is, any matrix representing ω has determinant ± 1) by Poincaré duality.

Poincaré duality identifies $H_2(M; \mathbb{Z})$ with $H^2(M; \mathbb{Z})$, and therefore, allows us to define ω as a pairing $H^2(M; \mathbb{Z}) \otimes H^2(M; \mathbb{Z}) \to \mathbb{Z}$. If $\alpha, \beta \in H^2(M; \mathbb{Z})$ and $[M] \in H_4(M; \mathbb{Z})$ is the fundamental class of M given by the orientation, we define

$$\omega(\alpha, \beta) = (\alpha \smile \beta)[M].$$

Here $\smile: H^2(M; \mathbb{Z}) \otimes H^2(M; \mathbb{Z}) \to H^4(M; \mathbb{Z})$ is the cup product which, in this case, is commutative.

When M has a smooth structure there is a third way to define ω by using the de Rham cohomology $H^*_{DR}(M)$. Let $\alpha, \beta \in H^2_{DR}(M)$ be represented by 2-forms which we also denote α, β. Then

$$\omega(\alpha, \beta) = \int_M \alpha \wedge \beta.$$

However we define it, ω is called the *intersection form* and is the basic invariant of a compact 4-manifold. In 1949 Whitehead [**W**] showed that the *homotopy* type of a compact, 1-connected four-manifold is determined by the isomorphism class of ω (cf. [**M2**]). This result was superseded in 1981 by Michael Freedman's classification of 4-manifolds by *homeomorphism* type [**Fr**].

EXAMPLES:

1. $M = S^4$. Since $H_2(S^4; \mathbb{Z}) = H^2(S^4; \mathbb{Z}) = 0$, $\omega = \emptyset$.
2. $M = S^2 \times S^2$. Then $H_2(S^2 \times S^2; \mathbb{Z})$ is generated by $a = S^2 \times pt$ and $b = pt \times S^2$. Here "pt" stands for a point of S^2. The matrix of ω in the basis $\{a, b\}$ is

$$\begin{pmatrix} 0 & 1 \\ 1 & 0 \end{pmatrix}.$$

3. $M = \mathbb{CP}^2$. Then $H_2(\mathbb{CP}^2; \mathbb{Z})$ has one generator, and the matrix of ω is (1).
4. $M = K = \{[z_0, z_1, z_2, z_3] \in \mathbb{CP}^3 : (z_0)^4 + (z_1)^4 + (z_2)^4 + (z_3)^4 = 0\}$ is the *Kummer surface*. By a calculation with characteristic classes [**M2**] it is possible to show that ω is represented by

$$-E_8 \oplus -E_8 \oplus 3 \begin{pmatrix} 0 & 1 \\ 1 & 0 \end{pmatrix},$$

where

$$E_8 = \begin{pmatrix} 2 & -1 & 0 & 0 & 0 & 0 & 0 & 0 \\ -1 & 2 & -1 & 0 & 0 & 0 & 0 & 0 \\ 0 & -1 & 2 & -1 & 0 & 0 & 0 & 0 \\ 0 & 0 & -1 & 2 & -1 & 0 & 0 & 0 \\ 0 & 0 & 0 & -1 & 2 & -1 & 0 & -1 \\ 0 & 0 & 0 & 0 & -1 & 2 & -1 & 0 \\ 0 & 0 & 0 & 0 & 0 & -1 & 2 & 0 \\ 0 & 0 & 0 & 0 & -1 & 0 & 0 & 2 \end{pmatrix}$$

Note that $\begin{pmatrix} 0 & 1 \\ 1 & 0 \end{pmatrix}$ is the intersection form for $S^2 \times S^2$, and E_8 is the Cartan matrix for the exceptional Lie algebra e_8.

It is important to know a little bit about unimodular symmetric bilinear forms ω over the integers. The *rank* $r(\omega)$ of ω is the dimension of the space on which ω is defined (in our case $\dim H_2(M; \mathbf{Z})$), and the *signature* $\sigma(\omega)$ is the number of positive "eigenvalues" minus the number of negative "eigenvalues" when ω is considered as a real form. The *type* of ω is defined by calling ω *even* if all diagonal entries are even (equivalently, $\omega(\alpha, \alpha) \in 2\mathbf{Z}$ for all $\alpha \in H^2(M; \mathbf{Z})$); otherwise ω is *odd*. Then the rank, signature, and type form a complete set of invariants for *indefinite* unimodular symmetric forms [M2], [HM, p. 25]. The classification of *positive definite* unimodular forms is much more difficult. At least there is the result due to Eisenstein and Hermite that there are finitely many isomorphism classes of positive definite unimodular forms of a given rank. Not all combinations of rank, signature, and type can occur; this applies to both definite and indefinite forms. The only nontrivial restriction is that if ω is even, then $\sigma(\omega)$ is divisible by 8. In this connection notice that $\sigma(E_8) = 8$. In fact, E_8 is the unique positive definite even form of rank 8. The proliferation of definite forms is evidenced by the following statistics: There are 2 even positive definite forms of rank 16, 24 or rank 24, and more than 10^{51} of rank 40!

The *Kirby–Siebenmann obstruction* $\alpha(M) \in \mathbf{Z}_2$ also appears in Freedman's classification. It is most easily described by stating that $\alpha(M) = 0$ if $M \times S^1$ is smoothable and $\alpha(M) = 1$ if $M \times S^1$ is not smoothable. ($M \times S^1$ is a 5-manifold, and smoothability is well understood in five dimensions.) Then we have the following

CLASSIFICATION (Freedman). *Compact, 1-connected topological 4-manifolds are in* $1 - 1$ *correspondence with pairs*

$$\{\langle \omega, \alpha \rangle : \textit{if } \omega \textit{ is even, then the relation } \frac{\sigma(\omega)}{8} \equiv \alpha(\!\!\mod 2) \textit{ is imposed}\}.$$

In particular, given a quadratic form ω, there is a 4-manifold $|\omega|$ realizing ω as its intersection form. If ω is even, the homeomorphism type of $|\omega|$ is uniquely determined by ω. For odd ω there are exactly two non-homeomorphic $|\omega|$ realizing ω. Incidentally, the uniqueness statement for the pair $\langle \emptyset, 0 \rangle$ is the four dimensional Poincaré conjecture in the topological category. Freedman's Theorem does not address the classification of differentiable 4-manifolds.

Differentiable 4-Manifolds.

Smooth 4-manifolds have resisted classification; dimension four seems particularly difficult for smoothness questions. In lower dimensions any topological manifold admits a unique smooth structure (up to diffeomorphism) [**Mo**], and in higher dimensions smooth structures correspond to reductions of the tangent bundle [**KS**]. A result of Rohlin in 1952 foreshadows the complications in dimension four.

THEOREM 1.2 (Rohlin [**R**]). *If a smooth, 1-connected, compact 4-manifold has even intersection form* ω, *then* $\sigma(\omega)$ *is divisible by 16.*

Note the contrast with the previous section; for $|\omega|$ to exist *topologically* if ω is even, it is enough to have $\sigma(\omega)$ divisible by 8. Hence it follows from Rohlin's Theorem and Freedman's classification that $|E_8|$ exists as a topological manifold, but not as a smooth one.

In some expositions of Rohlin's Theorem (see [**FK**], [**KM1**]) the hypothesis that ω is even is replaced by the hypothesis that M is spinable, i.e., that the second Steifel–Whitney class w_2 vanish. If M is simply connected these are equivalent: $w_2 = 0$ if and only if ω is even. We sketch the argument. Since $\pi_1(M) = 0$, we have $H_1(M; \mathbb{Z}_2) = 0$ (the latter can replace the former as the hypothesis), and the Universal Coefficient Theorem for cohomology gives an isomorphism $H^2(M; \mathbb{Z}_2) \xrightarrow{\sim} \mathrm{Hom}(H_2(M; \mathbb{Z}), \mathbb{Z}_2)$. Under this isomorphism $w_2 \in H^2(M; \mathbb{Z}_2)$ goes into the element of $\mathrm{Hom}(H_2(M; \mathbb{Z}), \mathbb{Z}_2)$ which, evaluated on an oriented surface $\Sigma \subseteq M$, is 0 or 1 according as $TM|_\Sigma$ is trivial or nontrivial. Now $TM|_\Sigma = \nu \oplus T\Sigma$ where ν is the normal bundle of Σ in M. Since Σ is an oriented surface, $T\Sigma$ is stably trivial. The obstruction to trivializing ν is the Euler class $\chi(\nu)$, which is the intersection number of Σ with itself. But since we consider $\nu \oplus T\Sigma$, we can reduce $\chi(\nu) \in H^2(\Sigma; \pi_1(SO(2))) = H^2(\Sigma; \mathbf{Z})$ to $\chi(\nu)_{\mathrm{mod}\ 2} \in H^2(\Sigma; \pi_1(SO(4))) = H^2(\Sigma; \mathbf{Z}_2)$. This is zero precisely when $\omega(\Sigma, \Sigma) \in 2\mathbf{Z}$.

Rohlin's Theorem implies that $|\omega|$ does not exist for a large set of even ω; Donaldson's Theorem takes care of the positive definite case.

THEOREM 1.3 (Donaldson). *$|\omega|$ does not exist smoothly for any nonzero even positive definite ω.*

In particular, $|E_8 \oplus E_8|$ does not exist smoothly. Recall from Example 4 of the previous section that the Kummer surface $K = \left| -E_8 \oplus -E_8 \oplus 3 \begin{pmatrix} 0 & 1 \\ 1 & 0 \end{pmatrix} \right|$ does exist smoothly. Since $S^2 \times S^2 = \left| \begin{pmatrix} 0 & 1 \\ 1 & 0 \end{pmatrix} \right|$ also exists smoothly (Example 2), one would like to do surgery on K to obtain a smooth $|E_8 \oplus E_8|$ by removing $\left| 3 \begin{pmatrix} 0 & 1 \\ 1 & 0 \end{pmatrix} \right|$ and reversing the orientation. In the next section we will see that out of the fact that this surgery is impossible pops the existence of $\mathbf{R}^4_{\mathrm{fake}}$!

We close this section with a recent result of Quinn which shows that all compact 4-manifolds are *almost smooth*, that is, smoothable in the complement of any point.

THEOREM 1.4 (Quinn [**Q**]). *Any noncompact 4-manifold is smoothable.*

A Surgical Failure.

Before we bring the Kummer surface into the *OR*, let's recall how to remove homology in a simpler case—amputating handles from a Riemann surface.

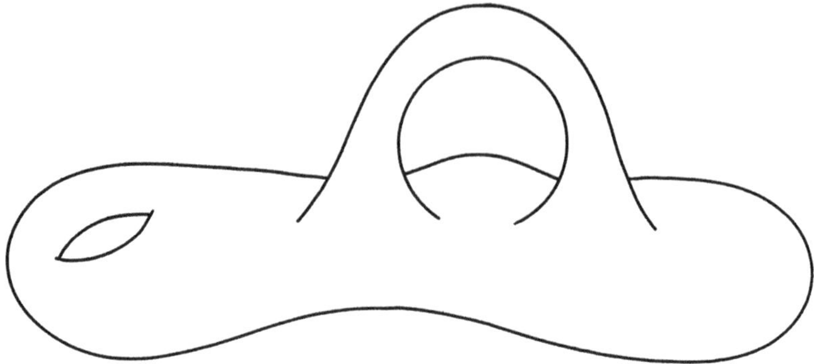

Consider the surface where the (antisymmetric) intersection form of the handle is $\begin{pmatrix} 0 & 1 \\ -1 & 0 \end{pmatrix}$. To remove this homology we implement one of three procedures:

1. Represent the homology by an embedding of a punctured torus $S^1 \times S^1 - D^2$. (The puncture D^2 is where the handle is joined to the surface.) Now crush the "wedge" $S^1 \vee S^1 = (S^1 \times pt) \cup (pt \times S^1) \subset S^1 \times S^1$ to a point. Then $(S^1 \times S^1)/(S^1 \vee S^1) = S^2$ since

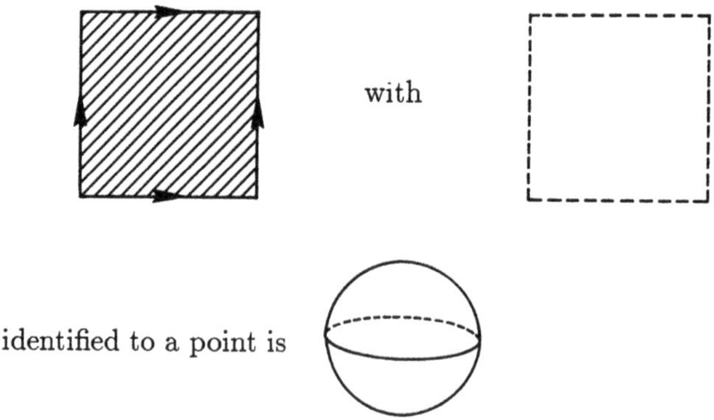

with

identified to a point is

2. Find an S^1 surrounding the handle.

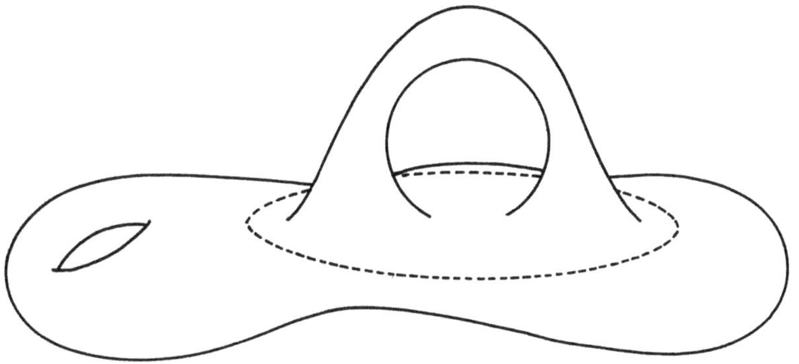

Cut out the handle

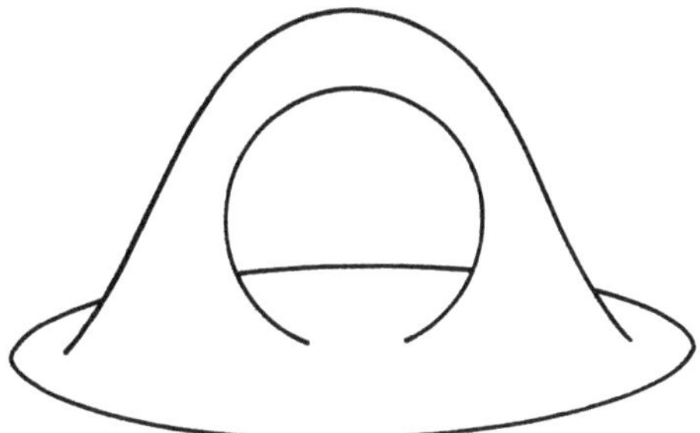

and glue back in a disk.

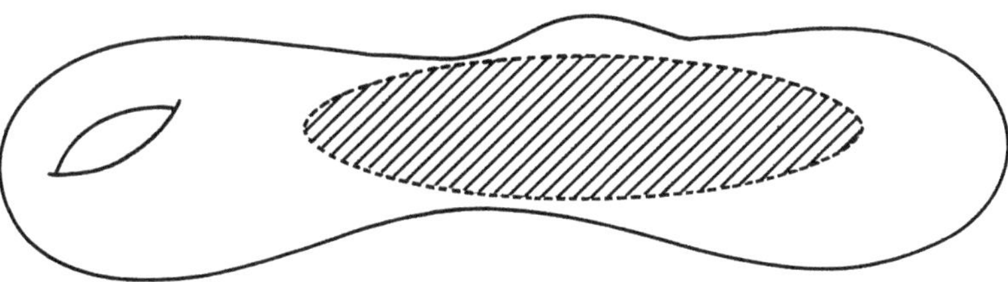

3. Cut out part of the handle

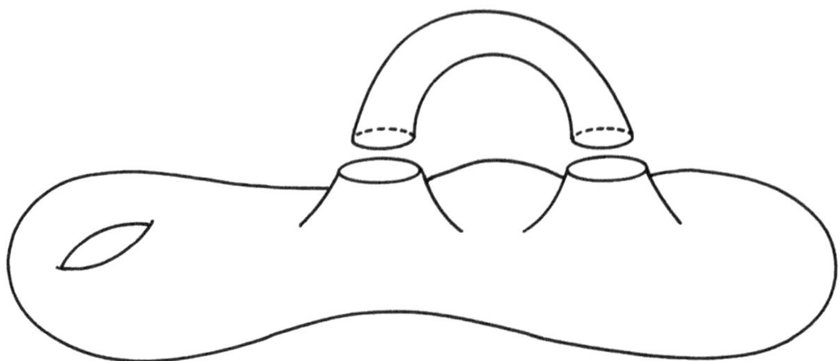

and glue back in 2 disks ($D^2 \times S^0$).

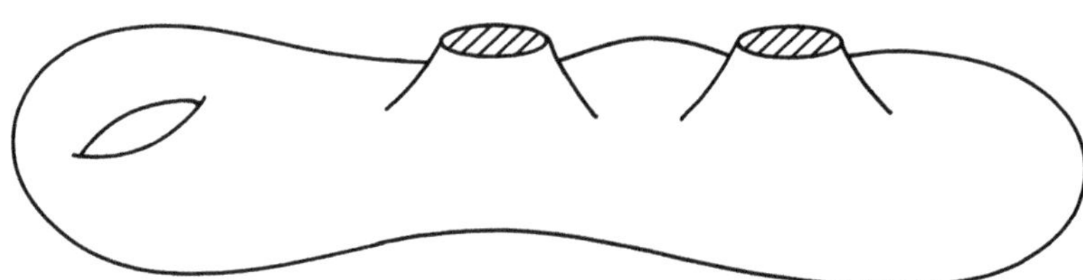

We represent the Kummer surface K by

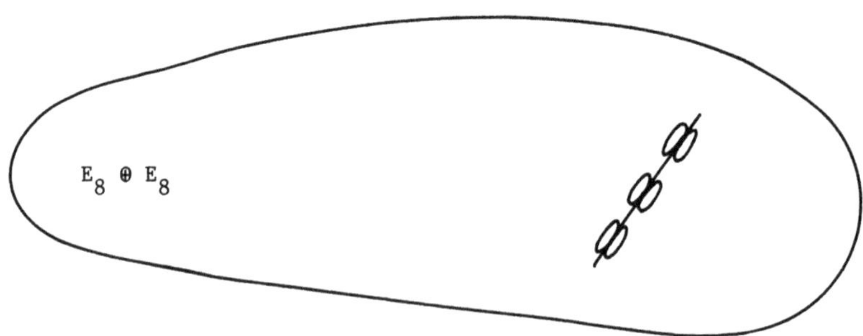

where

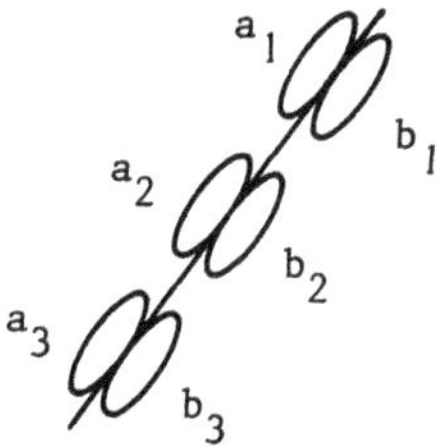

is a pictorial representation of the second homology whose generators are $a_i, b_i \in H_2(K, \mathbb{Z})$, $i = 1, 2, 3$. We then use the following topological results, both of which can be considered as corollaries of a result about *Casson handles* [**Fr**]. (See [**C**] for Casson's work.)

THEOREM 1.5. *The homology generated by a_i, b_i is represented by a collared topological embedding of $X = 3(S^2 \times S^2) - D^4$ in K. Furthermore, a smoothly equivalent embedding of X in $3(S^2 \times S^2)$ represents the homology of $3(S^2 \times S^2)$.*

THEOREM 1.6. *Any noncompact 4-manifold V without boundary which is simply connected, satisfies $H_2(V; \mathbb{Z}) = 0$, and has a single end homeomorphic to $S^3 \times [0, \infty)$ is homeomorphic to $\mathbb{R}^4$.*

Theorem 1.5 asserts the existence of the diagram

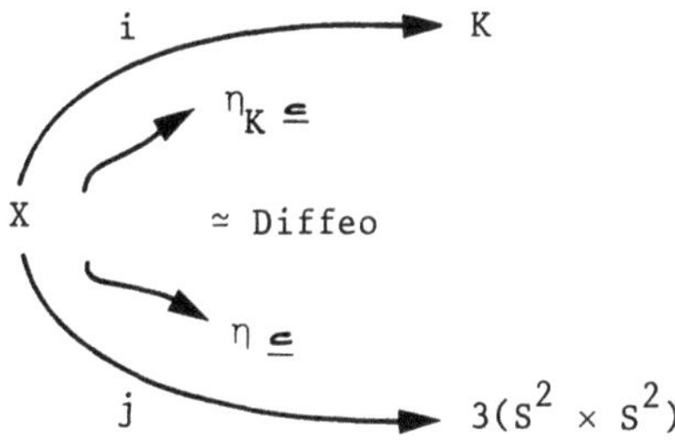

Here η_K and η are neighborhoods of the image of X in K and $3(S^2 \times S^2)$ respectively. Also, a *collared embedding* of X is one for which the image of ∂X has a product neighborhood.

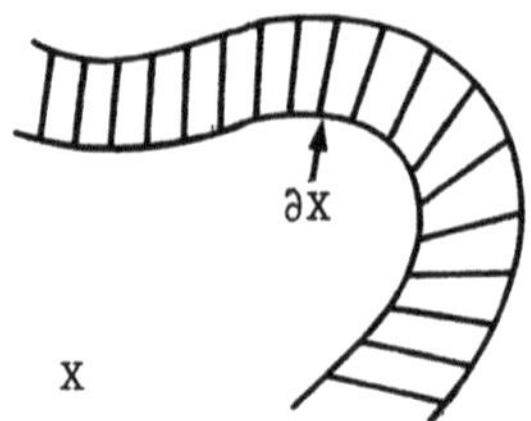

The collaring of $j(X) \subset 3(S^2 \times S^2)$ implies that $V = 3(S^2 \times S^2) - j(X)$ has a single end homeomorphic to $S^3 \times [0, \infty)$. So by Theorem 1.6, V is homeomorphic to $\mathbf{R}^4$.

Now we use Donaldson's Theorem to show that the differentiable structure V inherits from $3(S^2 \times S^2)$ is not the standard one. Suppose there is a *smoothly* embedded S^3

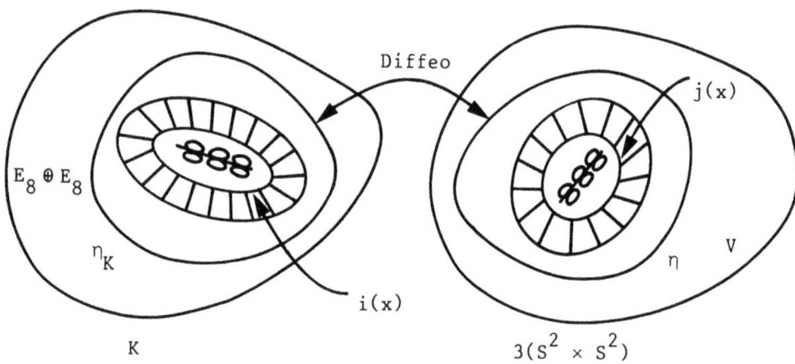

surrounding $i(X)$ in the collar of $i(X)$. Thus by the four dimensional analogue of the second surgical procedure outlined above, we could construct a smooth manifold $|E_8 \oplus E_8|$. But Donaldson's result forbids this. Thus there is no such S^3. By the diffeomorphism $\eta_K \simeq \eta$ there is no smooth S^3 in the collar of $j(X)$. Denoting the compact set $3(S^2 \times S^2) - (j(X) \cup \text{collar})$ by C, we see that in $V \overset{\text{top}}{\simeq} \mathbf{R}^4$ there is no smoothly embedded S^3 surrounding C. Then V satisfies the characterization (1.1), and we define $\mathbf{R}^4_{\text{fake}} = V$.

We close by mentioning two open problems. Our argument gives the existence of a new differentiable structure on $\mathbf{R}^4$. How many other exotic $\mathbf{R}^4$'s exist? Robert Gompf has constructed three distinct fake $\mathbf{R}^4$'s [G]. Finally, are there *any* fake S^4's? The smooth four dimensional Poincaré conjecture asserts that there are none. Freedman's work demonstrates that a manifold homotopy equivalent to S^4 is homeomorphic to S^4; the question posed here is whether a manifold homeomorphic to S^4 is diffeomorphic to S^4.

§2 THE YANG–MILLS EQUATIONS

From now on we deal with a certain orbit space $\mathcal{M}$. This "moduli space" is the set of solutions to the self-dual Yang–Mills equations, but divided out by a natural equivalence. The self-dual Yang–Mills equations are an elliptic (except along orbits of the group of gauge transformations) system of partial differential equations for a connection on a vector bundle over a smooth 4-manifold. The group of gauge transformations is the group of natural equivalences in the vector bundle, hence the natural equivalence in the problem, and so it acts on the space of solutions to the self-dual Yang–Mills equations.

Intuitively, one should view this moduli space as an analogue of the more familiar moduli spaces for Riemann surfaces of fixed genus. (The genus corresponds to the Chern number of the bundle in our problem.) Here one starts with the obviously infinite dimensional space of all possible surfaces of the same genus. After dividing out by conformal equivalence, the moduli space obtained is a finite dimensional cell complex, the Riemann moduli space. This Riemann space is a standard orbit space for a cell (the Teichmüller space) under the action of a discrete group (the modular group).

Our moduli space $\mathcal{M}$ is quite similar. As we said, it is the orbit space for solutions of the self-dual Yang–Mills equations under the action of gauge transformations. The Atiyah–Singer Theorem gives the "abstract dimension" of $\mathcal{M}$ as the index of a certain elliptic complex, which in our example is 5. This is the actual dimension of $\mathcal{M}$ if certain cohomology vanishes. The singularities resulting from nonvanishing cohomology have two sources: They arise from fixed points of the group action or result from the failure of the equations to be "onto". Both cases are treated more precisely in §3 and §4. Singularities of the first type correspond to reducible connections. Hence in this chapter, after introducing the moduli space $\mathcal{M}$, we study the special case of a $U(1)$ line bundle. At last we will have the background to outline the proof of Donaldson's Theorem. In succeeding chapters we will prove results about the topology of $\mathcal{M}$.

The Yang–Mills equations were introduced by physicists in the 50's to describe "strong interactions". A very readable account of this side of the story can be found in ['] (see also [**May**]). Although Yang–Mills is a hyperbolic system on Minkowski space, the elliptic versions of these equations are also relevant in physics; they appear in quantum field theories. We study the elliptic equations on Riemannian 4-manifolds. The reader may wish to read Ronald Stern's expository article [**Sn**] for more background and references. We also recommend [**A2**], [**AB**], [**AHS**], [**B**], [**BL**], [**DV**], [**JT**], and [**Pa**] for further material.

Connections.

A *principal bundle* P over a manifold M with structure group G is a twisted version of $M \times G$. (We will always have $G = U(1) = S^1$ or $G = SU(2)$.) Although the bundle P will appear from time to time, it is not a crucial ingredient in our exposition, and we will usually give an alternative phrasing that avoids its use. Thus we defer to standard texts (e.g. [**KN1**]) for the definition and elementary properties of principal bundles. Instead, when analysis is involved we find it much better to work with a *vector bundle* η associated to P. The vector bundle η is defined by specifying a representation $G \to \mathrm{Aut}(V)$ and forming the twisted product $\eta = P \times_G V$. We will need a local description of η. Namely, let $\{\mathcal{O}_\alpha\}$ be an open cover of M such that $\eta|_{\mathcal{O}_\alpha} \simeq \mathcal{O}_\alpha \times V$. We assume definite trivializations are chosen and that on the overlaps

$$s_{\beta\alpha} : \mathcal{O}_\alpha \cap \mathcal{O}_\beta \longrightarrow G \subseteq \mathrm{Aut}(V)$$

are the *transition functions*. In other words,

$$\eta = \coprod_\alpha \mathcal{O}_\alpha \times V / \sim,$$

where for $x \in \mathcal{O}_\alpha, y \in \mathcal{O}_\beta, v, w \in V$,

$$\langle x, v \rangle \sim \langle y, w \rangle \quad \text{iff} \quad x = y \quad \text{and} \quad w = s_{\beta\alpha}(v).$$

More details concerning this local construction can be found in Steenrod [**Ste**].

Although these trivializations allow us to identify different fibers locally, the identification is not canonical. To relate fibers intrinsically we must specify a *connection, covariant, derivative,* or *gauge potential*. (These are equivalent notions.) A covariant derivative D is a first order differential operator

$$D : C^\infty(\eta) \longrightarrow C^\infty(\eta \otimes T^*M).$$

Here $C^\infty(\eta)$ denotes the space of C^∞ sections of the bundle η. (We will usually denote $C^\infty(\eta \otimes \Lambda^i T^*M)$ by $\Omega^i(\eta)$.) If η has a metric (i.e., the bundle has structure group $U(n)$), then we require the "Leibnitz rule"

$$d(\sigma, \tau) = (D\sigma, \tau) + (\sigma, D\tau)$$

for $\sigma, \tau \in C^\infty(\eta)$. If, in addition, the group of the bundle is $SU(n)$, then we require the existence of orthonormal local sections $\sigma_1, \sigma_2, \ldots, \sigma_n$ for which

$$D(\sigma_1 \wedge \cdots \wedge \sigma_n) = 0.$$

There is also a definition of the connection in terms of the principal bundle P [**KN1**]. We will rely more on the local description of D. Namely, on $\mathcal{O}_\alpha$ the covariant derivative takes the form $d + A_\alpha$, where

$$A_\alpha : \mathcal{O}_\alpha \longrightarrow T^*(\mathcal{O}_\alpha) \otimes \mathfrak{g},$$

and $\mathfrak{g}$ is the Lie algebra of the gauge group. (In the $U(1)$ case $\mathfrak{g} = i\mathbf{R}$ and for $SU(2)$, $\mathfrak{g} = \mathfrak{su}(2)$, the algebra of traceless skew-Hermitian matrices.) The notation can get unwieldy since A carries three sets of indices—matrix indices, an index indicating the coordinate chart, and a 1-form index. So, for example, $A_{\alpha,i}(x)$ is a matrix valued function on $\mathcal{O}_\alpha$. Different local expressions for A are related by

$$(2.1) \qquad A_\alpha = s_{\beta\alpha}^{-1} ds_{\beta\alpha} + s_{\beta\alpha}^{-1} A_\beta s_{\beta\alpha}$$

on the overlap $\mathcal{O}_\alpha \cap \mathcal{O}_\beta$. Note that A has no global description on M. (However, the connection can be identified with an element of $C^\infty(\mathfrak{g} \otimes T^*P)$.) We denote the space of all connections by $\mathcal{A}$. It follows from (2.1) that the difference of two connections is in the linear space $\Omega^1(\operatorname{ad}\eta)$, whence $\mathcal{A}$ is an affine space. Here ad η is the bundle $P \times_G \mathfrak{g}$, where G acts on $\mathfrak{g}$ via the adjoint action. Alternatively, ad $\eta = \cup_{x \in M}(\operatorname{ad}\eta_x)$, where ad $\eta_x \subset \operatorname{End}(\eta_x)$ is the Lie algebra of endomorphisms of the fiber η_x which lie in $\mathfrak{g}$. Our local description reflects the affine structure in that a connection is expressed as a base connection d plus an arbitrary element $A_\alpha \in \Omega^1(\operatorname{ad}\eta|_{\mathcal{O}_\alpha})$.

30

Fix a covariant derivative D on η. Then we can tensor the de Rham complex

$$\Omega^0 \xrightarrow{\ d\ } \Omega^1 \xrightarrow{\ d\ } \Omega^2 \xrightarrow{\ d\ } \ldots \xrightarrow{\ d\ } \Omega^n$$

with η to obtain

$$(2.2) \qquad \Omega^0(\eta) \xrightarrow{\ D\ } \Omega^1(\eta) \xrightarrow{\ D\ } \Omega^2(\eta) \xrightarrow{\ D\ } \ldots \xrightarrow{\ D\ } \Omega^n(\eta).$$

The first operator is just the covariant derivative. This is extended to $\Omega^i(\eta)$ by defining

$$D(\sigma \otimes \theta) = D\sigma \wedge \theta + \sigma \otimes d\theta$$

for $\sigma \in C^\infty(\eta)$, $\theta \in \Omega^i$. The *curvature* (or *gauge field*) of the connection is determined by $D^2 : \Omega^0(\eta) \to \Omega^2(\eta)$. It is well-known that D^2 is a zeroth order operator, that is, multiplication by an element $F \in \Omega^2(\operatorname{ad} \eta)$. (This will follow from our local formulas below.) The curvature F is sometimes written

$$F = dA + A \wedge A,$$

the combination of exterior multiplication and matrix multiplication defining $A \wedge A$. More explicitly, if x^i are coordinates for M on $\mathcal{O}_\alpha$, and if $D_\alpha = d + A_\alpha$, then

$$F_\alpha = \frac{1}{2} \sum_{i,j} F_{\alpha,ij} \, dx^i \wedge dx^j,$$

where

$$(2.3) \qquad F_{\alpha,ij} = \frac{\partial A_{\alpha,j}}{\partial x^i} - \frac{\partial A_{\alpha,i}}{\partial x^j} + [A_{\alpha,i}, A_{\alpha,j}].$$

(Note that the matrix indices in the Lie algebra are suppressed; our indices are cotangent indices and coordinate indices.) It follows that, suppressing the 2-form indices on F,

$$(2.4) \qquad F_\alpha = s_{\beta\alpha}^{-1} F_\beta s_{\beta\alpha} = \operatorname{ad}(s_{\beta\alpha}^{-1}) F_\beta,$$

so that $F \in \Omega^2(\operatorname{ad} \eta)$ as claimed.

Suppose now that M is a four dimensional oriented Riemannian manifold. Then the (algebraic) *star operator* $* : \Omega^2 \to \Omega^2$ is defined fiberwise by

$$\theta \wedge *\phi = (\theta, \phi)\mathrm{vol}, \qquad \theta, \phi \in \Omega^2,$$

where (,) is the metric and vol is the volume form. Since $*^2 = 1$ we can decompose $\Omega^2 = \Omega^2_+ \oplus \Omega^2_-$ into the direct sum of the ± 1 eigenspace of $*$. This corresponds to the Lie algebra decomposition $\mathfrak{so}(4) = \mathfrak{so}(3) \oplus \mathfrak{so}(3)$. Elements of Ω^2_+ are *self-dual* and elements of Ω^2_- are *anti-self-dual*. This decomposition extends to $\Omega^2(\eta)$ for any bundle η. In particular, the curvature $F = F_+ + F_-$ decomposes into self-dual and anti-self-dual pieces.

Finally, we introduce the group of bundle automorphisms of P, which we refer to as the group of *gauge transformations*. A word of warning: Some mathematicians call *this* the "gauge group", a term usually reserved for G. A gauge transformation is a fiber preserving map $s : P \to P$ which satisfies $s(p \cdot g) = s(p) \cdot g$ for $p \in P$, $g \in G$. Equivalently, s is a cross section of the bundle of groups Aut $\eta = P \times_G G$, where G acts by conjugation. Aut η can also be described as $\cup_{x \in M}$ Aut η_x, where Aut $\eta_x \subseteq GL(\eta_x)$ is the Lie group automorphism of the fiber η_x which lie in G. We denote the group of gauge transformations by $\mathcal{G} = C^\infty(\text{Aut } \eta)$. Note that Aut η is canonically a bundle of groups, *not* a principal bundle. Aut η has a trivial subbundle $P \times_G Z$, where Z is the center of G. For $G = SU(2)$, $Z = \mathbb{Z}_2$ and $\mathbb{Z}_2 = C^\infty(P \times_G Z)$ is the center of $\mathcal{G}$. In general Aut η is not trivial, but Aut η is trivial if G is abelian. $\mathcal{G}$ is an infinite dimensional Lie group (the group structure is given by pointwise multiplication of sections) whose Lie algebra is $C^\infty(\text{ad } \eta)$. We discuss the manifold structure of $\mathcal{G}$ in §3.

Let $s \in \mathcal{G}$. Then s acts on the space $\mathcal{A}$ of covariant derivatives by pullback:

$$(2.5) \qquad\qquad s^*(D) = s^{-1} \circ D \circ s,$$

i.e., on sections $\sigma \in C^\infty(\eta)$,

$$s^*(D)\sigma = s^{-1} D(s\sigma).$$

In terms of our local description,

$$(2.6) \qquad\qquad s^*(A_\alpha) = s^{-1} ds + s^{-1} A_\alpha s,$$

where now $s : \mathcal{O}_\alpha \to G$. The induced action on curvature is

$$(2.7) \qquad\qquad s^*(F) = s^{-1} F s.$$

In (2.7) s acts trivially on the "form part" of $F \in \Omega^2(\text{ad } \eta)$. Other authors allow s to act on D by pushforward:

$$s \cdot D = s \circ D \circ s^{-1}.$$

This reverses the role of s and s^{-1} in (2.5), but makes no difference in the end.

Topological Quantum Numbers.

Mathematicians call them Chern classes. We recall that if ξ is a complex vector bundle with connection, and if F is the curvature, then

$$(2.8) \qquad c(\xi) = [\det(1 + \frac{i}{2\pi}F)] \in H^*(M; \mathbb{Z})$$

is the total Chern class [**KN2**]. (The brackets denote cohomology. For convenience we will no longer distinguish between a form and its cohomology class in this section.) In the case of a $U(1)$ complex line bundle λ, the curvature is an ordinary 2-form f, and (2.8) simplifies:

$$c(\lambda) = 1 + \frac{i}{2\pi}f,$$

and

$$c_1(\lambda) = \frac{i}{2\pi}f$$

classifies these bundles topologically. It is also true that every element of $H^2(M; \mathbb{Z})$ corresponds to a line bundle. (We discuss the proof in Appendix E.) If η is an $SU(2)$ 2-plane bundle, then

$$F = \begin{pmatrix} f_1^1 & f_2^1 \\ f_1^2 & f_2^2 \end{pmatrix} \in \mathfrak{su}(2),$$

with

$$f_1^1 + f_2^2 = 0.$$

Now (2.8) yields

$$c_1(\eta) = 0,$$

$$c_2(\eta) = \frac{1}{4\pi^2}\left(f_2^1 \wedge f_1^2 - f_1^1 \wedge f_2^2\right),$$

$$= \frac{1}{8\pi^2}tr(F \wedge F).$$

The characteristic class $c_2(\eta)$ classifies $SU(2)$ bundles over compact 4-manifolds, but this classification fails in higher dimensions [**DW**], [**Pe**]. (See Appendix E for the proof in four dimensions.) We can evaluate $c_2(\eta)$ on the fundamental class $[M]$ of a 4-dimensional oriented manifold to obtain the *topological charge*

$$(2.9) \qquad k = -c_2(\eta)[M] = \frac{-1}{8\pi^2}\int_M tr(F \wedge F).$$

The convention $k = -c_2(\eta)[M]$ is firmly entrenched in the Yang–Mills literature, and we are powerless to correct the sign. Sometimes the first Pontrjagin class $p_1(\eta)$ is used instead of $c_2(\eta)$. The formula $p_1(\eta) = -2c_2(\eta)$ gives the relationship to c_2 (assuming the group of η is $SU(2)$).

When can an $SU(2)$ bundle η be written as the direct sum of line bundles $\lambda_1 \oplus \lambda_2$? If η decomposes, the Whitney product formula implies

$$0 = c_1(\eta) = c_1(\lambda_1) + c_1(\lambda_2),$$

$$c_2(\eta) = c_1(\lambda_1) \smile c_1(\lambda_2),$$

$$= -c_1(\lambda_1) \smile c_1(\lambda_1).$$

Thus

$$(2.10) \qquad\qquad k = \omega(\alpha, \alpha),$$

where $\alpha = \pm c_1(\lambda_1)$ and ω is the intersection form. This proves the "only if" half of

PROPOSITION 2.11. *The $SU(2)$ bundle η over M with second Chern number k splits topologically if and only if (2.10) holds for some α. For a fixed k the number of splittings is half the number of solutions to (2.10).*

In particular, if ω is positive definite, then no bundle η with negative k splits.

PROOF: Suppose $\alpha \in H^2(M; \mathbb{Z})$ satisfies $\omega(\alpha, \alpha) = k$. Then by the Bundle Classification Theorem $(E.5)$, there is a complex line bundle λ with $c_1(\lambda) = \alpha$. Furthermore, the preceeding calculation shows that $-c_2(\lambda \oplus \lambda^{-1})[M] = k$, and by $(E.5)$, $\eta \simeq \lambda \oplus \lambda^{-1}$ as $SU(2)$ bundles.

In this discussion we have implicitly assumed that $H^2(M; \mathbb{Z})$ is torsion free. Otherwise, we must divide out the torsion to define the intersection form ω. Note that Tor $H^2(M; \mathbb{Z}) = 0$ if $H_1(M; \mathbb{Z}) = 0$ (cf. $(E.1)$).

The Yang–Mills Functional.

Let M be a four dimensional Riemannian manifold and η a vector bundle over M. For convenience of exposition we assume that the group of η is $SU(2)$. The inner product

$$(2.12) \qquad (A, B) = -tr(B), \qquad A, B \in \mathfrak{su}(2),$$

together with the Riemannian structure of M makes ad $\eta \otimes \Lambda^2 T^* M$ a Riemannian bundle. We define the *Yang-Mills functional*

$$(2.13) \qquad \mathcal{YM}(D) = \int_M |F_D|^2 * 1, \qquad D \in \mathcal{A},$$

where $F = F_D$ is the curvature of D, and $*1$ is the volume form of the metric. $\mathcal{YM}(D)$ measures the "strength" of the gauge field F. The *Yang–Mills equations* are the Euler–Lagrange equations for the action integral (2.13). Since

$$\begin{aligned}
F_{D+tA}\sigma &= (D + tA)(D + tA)\sigma \\
&= D^2\sigma + t(D(A\sigma) + A \wedge D\sigma) + t^2(A \wedge A)\sigma \\
&= [F + t(DA) + t^2(A \wedge A)]\sigma,
\end{aligned}$$

we have

$$\begin{aligned}
\frac{d}{dt}\mathcal{YM}(D + tA)|_{t=0} &= \frac{d}{dt}|_{t=0} \int_M |F + tDA + t^2 A \wedge A|^2 * 1, \\
(2.14) \qquad\qquad &= 2 \int_M (DA, F) * 1, \\
&= 2 \int_M (A, D^* F) * 1.
\end{aligned}$$

Here $D^* : \Omega^2(\text{ad } \eta) \to \Omega^1(\text{ad } \eta)$ is the (formal) adjoint of $D : \Omega^1(\text{ad } \eta) \to \Omega^2(\text{ad } \eta)$. In terms of the $*$ operator,

$$D^* = - * D * .$$

The Euler–Lagrange equations are obtained by setting the first variation (2.14) equal to zero for all A. The result is

$$(2.15) \qquad\qquad D^* F_D = 0 = D * F_D.$$

We note that

$$(2.16) \qquad\qquad D F_D = 0$$

is the Bianchi identity which always holds. If F satisfies (2.15), then F is a *Yang-Mills field* and D a *Yang-Mills connection*. Equations (2.15) and (2.16) explain why Yang–Mills is considered to be a nonlinear version of Hodge theory; replacing D by d and F_D by f we obtain the equations for a harmonic 2-form f. The nonlinearity of the equations reflects the noncommutativity of G. If $g_{ij} = \delta_{ij}$ is the flat metric on the base, (2.15) becomes

$$(D^*F)_j = \sum_i \frac{\partial F_{ij}}{dx^i} + [A_i, F_{ij}] = 0$$

in local coordinates.

Equation (2.15) is *not* elliptic, basically due to the presence of a large symmetry group. In fact, (2.7) and the ad-invariance of (2.12) show that the functional $\mathcal{YM}$, and hence the Euler–Lagrange equations (2.15), is invariant under the action of gauge transformations. In other words, $\mathcal{YM}$ is a well-defined functional on the quotient space $\mathcal{A}/\mathcal{G}$. We note that for computation it is convenient to take a cross section of $\mathcal{A} \to \mathcal{A}/\mathcal{G}$. Cross sections exist locally, but not globally [S]. If we had global cross sections, many of our arguments would simplify. However, due to topological obstructions these do not exist. This fact is known to physicists as the Gribov ambiguity. We construct local cross sections in §3. It is tangent to these local slices that (2.15) is an elliptic system.

Decompose the curvature

$$F = F_+ + F_-$$

into its self-dual and anti-self-dual components. Then since Ω_+^2 is orthogonal to Ω_-^2,

$$(2.17) \qquad \mathcal{YM}(D) = \int_M |F_+|^2 + |F_-|^2.$$

Now

$$tr(F \wedge F) = tr(F_+ \wedge F_+) + tr(F_- \wedge F_-)$$
$$= tr(F_+ \wedge *F_+) - tr(F_- \wedge *F_-)$$
$$= -|F_+|^2 + |F_-|^2.$$

So by (2.9),

$$(2.18) \qquad 8\pi^2 k = \int_M |F_+|^2 - |F_-|^2.$$

Hence if $k > 0$,

$$\mathcal{YM} \geq 8\pi^2 k$$

with equality iff $F_- = 0$; i.e., F is self-dual. Similarly, if $k < 0$,

$$\mathcal{YM} \geq -8\pi^2 k$$

with equality iff $F_+ = 0$; i.e., F is anti-self-dual. Furthermore, if F is self-dual or anti-self-dual, then (2.15) follows from (2.16) so that F is automatically a solution to the Yang–Mills equations. In cases where the topological bound is the minimum, then, the second order equations (2.15) reduce to the first order *self-dual (anti-self-dual) Yang–Mills equation*

$$(2.19) \qquad F = \pm * F$$

if we seek absolute minima. However, there is no guarantee that in general this topological lower bound is attained. In some situations (e.g. line bundles over $S^2 \times S^2$) there are minima which are neither self-dual nor anti-self-dual.

We are interested in the case $k = 1$ and therefore in self-dual Yang–Mills fields, or *instantons*. Since the space of instantons is invariant under gauge transformations; we divide out the action of $\mathcal{G}$ to obtain the *moduli space*

$$\mathcal{M} = \{D \in \mathcal{A} : F_D = *F_D\}/\mathcal{G}.$$

$\mathcal{M}$ will be our principal object of study.

Line Bundles.

When $G = U(1)$ and M is a Lorentz manifold, then it is well known that (2.15) and (2.16) are Maxwell's equations [**MTW**]. We study positive definite metrics, hence elliptic versions of the equations. In fact, since $U(1)$ is abelian, for any $U(1)$ line bundle λ, ad λ is the trivial bundle $M \times i\mathbb{R}$. So the curvature f of a connection d' is an ordinary 2-form. Equations (2.15) and (2.16) reduce to

$$df = 0,$$

$$d^* f = 0,$$

demonstrating that the elliptic version of Maxwell's equations is just the Hodge–de Rham equations for harmonic 2-forms. The Hodge Theorem implies

THEOREM 2.20. *If λ is a line bundle, then the curvature of any Yang–Mills connection d' is the unique harmonic two-form f representing $c_1(\lambda)$.*

Of course, the Yang–Mills connection d' in (2.20) is not unique: Remember the gauge transformations! Assume that M is simply connected so that every gauge transformation on λ can be written $s = e^{iu}$ for some function u. Let d_0 be a fixed Yang–Mills connection on λ. Then any other Yang–Mills connection has the form

$$d' = d_0 + ia, \qquad a \in \Omega^1,$$

where $da = 0$. But by (2.6), the gauge transformation s pulls d' back to $d_0 + idu$. Since $H^1(M,\mathbf{R}) = 0$ by assumption, we can choose u so that $du = a$. Thus the gauge transformations act transitively on the space of Yang–Mills connections, and the moduli space for the full Yang–Mills equations (2.15)–(2.16) on λ is a point. We remark that if $\pi_1(M) \neq 0$, then the moduli space is the torus $H^1(M,\mathbf{R})/H^1(M,\mathbf{Z})$.

To apply this to the $SU(2)$ bundle case we need the following important

LEMMA 2.21. *The intersection form ω of M is positive definite if and only if there are no anti-self-dual harmonic 2-forms on M.*

PROOF: Let $f = f_+ + f_-$ be the decomposition of a harmonic 2-form into its self-dual and anti-self-dual parts. The lemma follows easily from the formula

$$\|f\|^2 = \omega(f_+, f_+) - \omega(f_-, f_-),$$

from which

$$\|f_-\|^2 = -\omega(f_-, f_-).$$

Under our basic topological assumption, then, we can describe completely the "split instantons" for split bundles (cf. Proposition 2.11).

PROPOSITION 2.22. *Suppose M has positive definite intersection form. Then for split $SU(2)$ bundles $\eta = \lambda \oplus \lambda^{-1}$ there is a unique self-dual Yang–Mills field*

$$F = \begin{pmatrix} f & 0 \\ 0 & -f \end{pmatrix}$$

which respects the splitting.

PROOF: Clearly, Yang–Mills connections on λ and λ^{-1} induce a Yang–Mills connection on $\lambda \oplus \lambda^{-1}$, and by (2.21) the curvature is self-dual. Conversely, a split self-dual connection on $\lambda \oplus \lambda^{-1}$ is self-dual when restricted to each line bundle. These connections are gauge equivalent by split gauge transformation (cf. the discussion following (2.20)).

Donaldson's Theorem.

We outline the main argument in Donaldson's work. Our setup is this: η is the $k = 1$ $SU(2)$ bundle over a compact, simply connected, oriented smooth 4-manifold M whose intersection form ω is positive definite. We list five topological results concerning the moduli space $\mathcal{M}$, each of which will be treated in later chapters.

I. *Let m be half the number of solutions to $\omega(\alpha, \alpha) = 1$. Then for almost all metrics on M, there exist $p_1, \ldots, p_m \in \mathcal{M}$ such that $\mathcal{M} - \{p_1, \ldots, p_m\}$ is a smooth 5-manifold. The points p_i are in $1 - 1$ correspondence with topological splittings $\eta = \lambda \oplus \lambda^{-1}$.*

II. *There exist neighborhoods $\mathcal{O}_{p_i}$ of p_i so that $\mathcal{O}_{p_i} \simeq$ cone on $\mathbf{CP}^2$.*

III. *$\mathcal{M}$ is orientable.*

IV. *$\mathcal{M} - \{p_1, \ldots, p_m\}$ is non-empty. In fact, there exists a collar $(0, \lambda_o] \times M \subset \mathcal{M}$, and $\overline{\mathcal{M}} = \mathcal{M} \cup M \supset [0, \lambda_o] \times M$ is a smooth manifold with boundary.*

V. *$\overline{\mathcal{M}}$ is compact.*

The topology of the moduli space is encoded in the following picture.

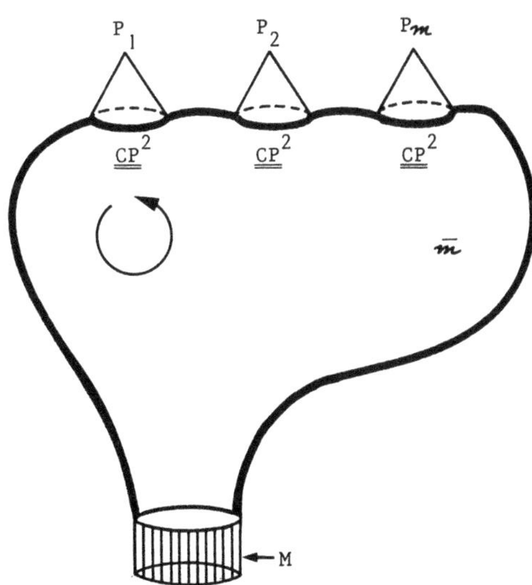

The heavy outline symbolizes compactness (V) and the arrow denotes orientability (III). We remark that it is unknown whether or not $\overline{\mathcal{M}}$ is connected.

As a first consequence of I–V,

LEMMA 2.23. *M is oriented cobordant to the disjoint union* $\underbrace{\pm \mathbf{CP}^2 \amalg \cdots \amalg \pm \mathbf{CP}^2}_{m}$.

PROOF: $\overline{\mathcal{M}} - \cup_i \mathcal{O}_{p_i}$ is the oriented cobordism. (The reader will find a brief discussion of cobordism in Appendix B.)

At this stage, we make no claim about the orientations of the $\mathbf{CP}^2$'s in (2.23). For example, if some singular points happened to be in a different component of $\overline{\mathcal{M}}$, then not all of the $\mathbf{CP}^2$'s would be positively oriented. However, it will be clear from the proof of (2.25) that all of the $\mathbf{CP}^2$'s are in the component of the collar and are positively oriented.

LEMMA 2.24. *Let ω be a positive definite symmetric unimodular form of rank $r = r(\omega)$, and let m be half the number of solutions α to $\omega(\alpha, \alpha) = 1$. Then $m \leq r$ with equality if and only if ω is diagonalizable over the integers.*

PROOF: A solution α gives a splitting $\mathbf{Z}^r = \mathbf{Z} \cdot \alpha \oplus \alpha^{\perp}$, orthogonal with respect to ω. If $\beta \neq \pm\alpha$ is another solution, $\beta \in \alpha^{\perp}$. The lemma now follows by induction on r.

THEOREM 2.25 (Donaldson [**D**]). *Let M be a compact, simply connected, oriented smooth 4-manifold with positive definite intersection form ω. Then*

$$\omega \simeq \underbrace{(1) \oplus \cdots \oplus (1)}_{m}$$

over the integers.

Of course, by simple linear algebra, ω is diagonalizable over $\mathbf{R}$. Donaldson's Theorem asserts that ω is also diagonalizable over $\mathbf{Z}$.

We can replace simple connectivity in Theorem 2.25 by the weaker hypothesis that there are no nontrivial homomorphisms of $\pi_1(M)$ into $SU(2)$. The latter condition implies that every flat $SU(2)$ bundle over M is trivial [**KN1**], and this is all we need. Then $H_1(M)$, the abelianization of $\pi_1(M)$, vanishes since any abelian group has a representation in $S^1 \subset SU(2)$. We summarize the various topological hypotheses in the diagram

$$\{\pi_1(M) = 0\} \Rightarrow \{\pi_1(M) \to SU(2) \text{ is trivial}\} \Rightarrow \{H_1(M) = 0\}.$$

No finite simple nonabelian group represents nontrivially in $SU(2)$ [**Do**, §26], and our proof of Donaldson's theorem allows all of these fundamental groups. However, many important groups (e.g. $\mathbf{Z}_2$) are excluded.

PROOF OF THEOREM 2.25: Since the signature of the intersection form is an oriented cobordism invariant [**Sto**, p. 219], it follows from (2.23) and the definiteness of ω that

$$r(\omega) = \sigma(\omega) \leq m \cdot \sigma(\mathbf{CP}^2) = m.$$

(Inequality would occur if some $\mathbf{CP}^2$ were glued on "backwards".) But $m \leq r(\omega)$ by (2.24). Hence $m = r(\omega)$, and ω is diagonalizable.

COROLLARY 2.26. *If $\omega \neq \emptyset$ and $\omega(\alpha, \alpha) \neq 1$ for all $\alpha \in H^2(M, \mathbf{Z})$, then M is not smoothable. This holds in particular if ω is even.*

There are many odd positive definite unimodular forms which do not take the value one [**CS**].

PROOF: The hypotheses imply that $m = 0$ so that M is cobordant to zero. Thus $\sigma(\omega) = 0$, and since ω is positive definite, $\omega = \emptyset$. This contradiction— a contradiction to the existence of a differentiable structure, and hence of Yang–Mills equations, moduli spaces, ... —shows that M carries no smooth structure.

Corollary 2.26 is the version of Donaldson's Theorem stated in §1. In particular, (2.26) shows that $|E_8 \oplus E_8|$ does not exist smoothly.

Finally, combining Donaldson's Theorem and Freedman's Classification, we obtain

COROLLARY 2.27. *Let M be a smooth simply connected 4-manifold. If $\omega(M)$ is even positive definite, then M is homeomorphic to S^4. If $\omega(M)$ is odd positive definite, then M is homeomorphic to a connected sum of positively oriented $\mathbf{CP}^2$'s.*

Any odd indefinite form is diagonalizable over the integers [**HM**], so that by Freedman's Theorem a smooth M with $\omega(M)$ odd indefinite is homeomorphic to II $\pm$ $\mathbf{CP}^2$. An even indefinite form is equivalent to $aE_8 \oplus b \begin{pmatrix} 0 & 1 \\ 1 & 0 \end{pmatrix}$ for some $a, b \in \mathbf{Z}$, $b \neq 0$. Recall that by Rohlin's theorem, $\left| aE_8 \oplus b \begin{pmatrix} 0 & 1 \\ 1 & 0 \end{pmatrix} \right|$ is not smoothable if a is odd. An outstanding problem in four dimensional smoothing theory, then, is to determine whether

$\left| 2E_8 \oplus b \begin{pmatrix} 0 & 1 \\ 1 & 0 \end{pmatrix} \right|$ is smoothable for $b = 1, 2$. (The Kummer surface realizes $b = 3$, and for $b > 3$ take the connected sum with $b - 3$ copies of $S^2 \times S^2$.)

The reader may wonder what difficulties arise as various hypotheses in Donaldson's Theorem are altered. We have already remarked that the simple connectivity requirement can be relaxed by assuming that $\pi_1(M)$ has no nontrivial representation in $SU(2)$. This condition arises because homomorphisms $\pi_1(M) \to SU(2)$ classify geometrically flat bundles, i.e., possibly nontrivial flat connections on topologically trivial $SU(2)$ bundles. The boundary of the compactified moduli space $\overline{\mathcal{M}}$, which in our case is merely a copy of M, arises by gluing standard instantons (over S^4) onto such flat bundles. If there are nontrivial flat bundles, then one expects a more complicated boundary, similar to that described by Taubes [T2] for the case when the intersection form is indefinite.

Many factors influence the dimension of the moduli space, which for any principal bundle P with compact structure group G over a compact 4-manifold M is given by the explicit formula [AHS] (note $p_1(\mathrm{ad}\, P)_{\mathbb{C}} = 2p_1(\mathrm{ad}\, P)$)

$$(2.28) \qquad \dim \mathcal{M}_P = p_1((\mathrm{ad}\ P)_{\mathbb{C}})[M] - (\dim G)(1 - b_1 + b_2^-).$$

Here $(\mathrm{ad}\ P)_{\mathbb{C}}$ is the complexified adjoint bundle of Lie algebras, p_1 denotes the first Pontrjagin class, b_1 is the first Betti number of M, and b_2^- is the dimension of the maximal subspace of $H^2(M; \mathbb{R})$ on which the intersection form is negative definite. If $\dim G > 3$ we run into real complications in the structure of the moduli space at reducible connections, so by default we restrict to $G = SU(2)$ or $SO(3)$. For $SU(2)$ bundles (2.28) reduces to

$$\dim \mathcal{M}_{SU(2)} = 8k - 3(1 - b_1 + b_2^-),$$

where, as always, k is minus the Chern number of the associated complex 2-plane bundle. Recall that $k \geq 0$ if $\mathcal{M}_{SU(2)}$ is nonempty, and if $k = 0$ all self-dual connections are flat. The case $k = 1$, $b_1 = b_2^- = 0$ yields the five dimensional moduli space used to prove Donaldson's Theorem. Any modification to k or b_1 would increase the dimension of $\mathcal{M}_{SU(2)}$, thereby excluding our cobordism argument, unless $b_2^- > 0$ also. But in the latter case, a construction of Taubes [T2] alluded to above indicates that the boundary is quite complicated. For $SO(3)$ bundles

$$(2.29) \qquad \dim \mathcal{M}_{SO(3)} = 2\ell - 3(1 - b_1 + b_2^-),$$

where ℓ is the Pontrjagin number of the associated real 3-plane bundle. Note that for $b_1 = b_2^- = 0$ and $\ell = 2$ the moduli space is one dimensional. Fintushel and Stern [**FS**] construct an $\ell = 2$ $SO(3)$ bundle ξ with $w_2(\xi) \neq 0$ directly from the intersection form ω. This requires that solutions $\alpha \in H^2(M; \mathbf{Z})$ to the equation $\omega(\alpha, \alpha) = 2$ exist. (Solutions do *not* exist for many positive definite ω—e.g., the 24 dimensional Leech lattice.) In addition, suppose that α is not the orthogonal direct sum of two vectors of length one. In this situation Fintushel and Stern prove that $\mathcal{M}_\xi$ is a compact one dimensional manifold with a single boundary point. This contradiction shows that the contemplated M is not smoothable, and so reproves Donaldson's Theorem for many intersection pairings (including $E_8 \oplus E_8$). Their proof avoids orientability, Taubes' Theorem, and the Collar Theorem, although they still need the hard analysis involved in compactness arguments. Furthermore, they obtain this restricted form of Donaldson's Theorem for almost any finite fundamental group. Also, this technique gives a new invariant for homology 3-spheres. We describe some of their results in §10.

The most promising direction for future investigation is 4-manifolds with boundary or with point singularities (which can be conformally blown up to infinity as in §7). However, index computations are anything but straightforward here [**APS**], and presently this limits the practicality of this application.

Our object is to show that $\hat{\mathcal{M}}$, the moduli space of irreducible, self-dual connections modulo gauge equivalence, is a smooth manifold.

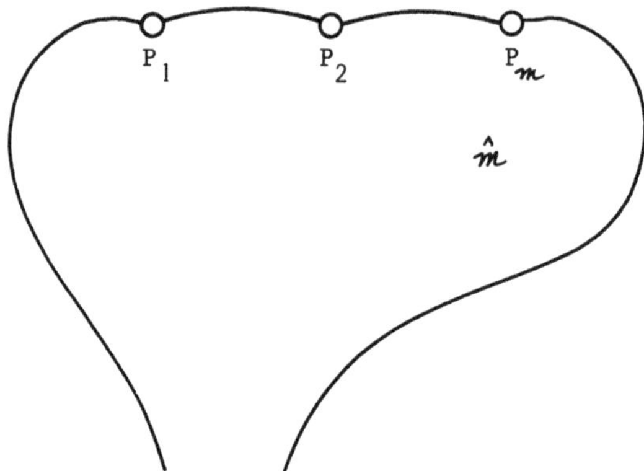

(Throughout, the hat "^" over a space of connections denotes the subspace of *irreducible* connections.) Since the proof is quite long, we give an informal sketch of the main ideas here. Let $\mathcal{X} = \mathcal{A}/\mathcal{G}$ be the orbit space of all connections under the action of gauge transformations. The first step is to construct a slice of this action away from reducible connections, thereby proving that $\hat{\mathcal{X}}$ is a manifold.

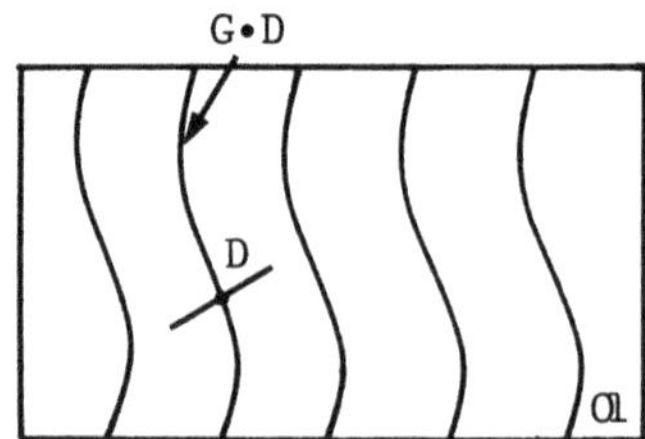

Now it is *not* known to be true for every metric g on our base manifold M that the moduli space $\hat{\mathcal{M}}_g \subseteq \mathcal{X}$ is a manifold. However, we show that it is true for almost all metrics. Thus we parametrize the space of metrics $\mathcal{C}$ and consider the set

$$\mathcal{S}\hat{\mathcal{D}} \subseteq \hat{\mathcal{A}} \times \mathcal{C},$$

where

$$\mathcal{S}\hat{\mathcal{D}} = \{\langle D, g \rangle : F_D \text{ is self-dual in the metric } g\}.$$

We show that $\mathcal{S}\hat{\mathcal{D}}$ is a manifold. We can mod out by gauge transformations as before (there is still a slice of the action). Hence $\mathcal{S}\hat{\mathcal{D}}/\mathcal{G}$ is also a manifold. Write this parametrized moduli space as

$$\mathcal{S}\hat{\mathcal{D}}/\mathcal{G} \simeq \bigcup_{g \in \mathcal{C}} \hat{\mathcal{M}}_g,$$

where $\hat{\mathcal{M}}_g$ is the moduli space for the metric g. Finally, we show that $\hat{\mathcal{M}}_g$ is a smooth manifold for almost all g by applying the Sard–Smale Theorem (that is, an infinite dimensional version of Sard's Theorem) to the projection

$$\mathcal{S}\hat{\mathcal{D}}/\mathcal{G} \simeq \bigcup_{g \in \mathcal{C}} \hat{\mathcal{M}}_g \to \mathcal{C}.$$

C^∞ spaces do not work for many of our arguments; they are not Banach spaces, and elliptic operators do not invert on them. So we replace C^∞ by either C^k or by Sobolev spaces. The relevant facts about Sobolev spaces are stated in the first section. Next, criteria for detecting reducible connections are given. After these preliminaries we turn to the proof of the main theorem as outlined above.

Sobolev Spaces.

We give a very brief account; see [**Au**], [**P**], [**Ma**, §3], and §6 for more detail. Let $\pi : \xi \to M$ be a Riemannian (or Hermitian) *vector* bundle with connection D over a compact n-dimensional Riemannian manifold. Then for each nonnegative integer ℓ we denote by $H_\ell(\xi)$ the space of sections whose derivatives of order $\leq \ell$ are square integrable. Thus $H_\ell(\xi)$ is the Hilbert space completion of $C^\infty(\xi)$ with respect to the inner product

$$(\sigma, \tau)_\ell = \int_M \sum_{i \leq \ell} (D^i \sigma, D^i \tau) * 1,$$

where $(D^i\sigma, D^i\tau)$ is computed using the inner product on $\xi \otimes \mathrm{Sym}^i(T^*M)$. We recall that $H_\ell(\xi) \subset C^k(\xi)$ for $\ell - \frac{n}{2} > k$ (Sobolev), and $H_\ell(\xi) \subset H_k(\xi)$ compactly for $\ell > k$ (Rellich). In nonlinear problems we deal with a *fiber bundle* $\pi : E \to M$. For example, the space $\mathrm{Map}(M, N)$ of maps between two manifolds is the space of sections of the trivial bundle $\pi : M \times N \to M$. That nonlinear fiber bundles enter into Yang–Mills is clear; the gauge transformations are sections of a bundle whose fibers are groups. In these cases $H_\ell(E)$ is a *Hilbert manifold* if $\ell > n/2$. (For our case $n = 4$ we require $\ell > 2$.) Although the manifold structure is difficult to describe, we can at least determine the tangent space to a section $s \in H_\ell(E)$. Suppose that $\gamma : (-1, 1) \to H_\ell(E)$ is a curve with $\gamma(0) = s$. Then for each $x \in M$, $\gamma_x(t) = \gamma(t)(x)$ is a curve in $E_x = \pi^{-1}(x)$. Thus $\gamma_x'(0) \in T_{\gamma_x(0)}(\pi^{-1}(x))$ and $\gamma'(0)$ is a section of $s^*(VTE)$, where VTE is the vertical tangent bundle of E. Therefore, $T_s(H_\ell(E)) = H_\ell(s^*(VTE))$.

We apply the foregoing to our situation: η is the $k = 1$ $SU(2)$ vector bundle over a compact 4-dimensional Riemannian manifold M. $\mathcal{G}$ is the group of gauge transformations, and $\mathcal{A}$ is the space of connections on η. $\mathrm{Aut}\,\eta$ and $\mathrm{ad}\,\eta$ are defined as in §2. Fix $\ell > 2$. We state the following, and defer the proofs to Appendix A.

1. Define the Hilbert manifold $\mathcal{G}_\ell = H_\ell(\mathrm{Aut}\,\eta)$. Recall that $\mathcal{G} = C^\infty(\mathrm{Aut}\,\eta)$ is the group of gauge transformations. It can be shown that $\mathcal{G}_\ell$ is a *Hilbert Lie group*, that is, an infinite dimensional Lie group modeled on a Hilbert space (cf. $(A.2)$). Since the Lie algebra of $\mathrm{Aut}\,\eta_x$ is $\mathrm{ad}\,\eta_x = T_{id}(\mathrm{Aut}\,\eta_x) = VT_{\langle x,id\rangle}(\mathrm{Aut}\,\eta)$, the Lie algebra of $\mathcal{G}_\ell$ is $T_{id}(\mathcal{G}_\ell) = H_\ell(id^*(VT(\mathrm{Aut}\,\eta))) = H_\ell(\mathrm{ad}\,\eta)$.

2. Choose a base connection $D_0 \in \mathcal{A}$. Define the *affine* space

$$\mathcal{A}_{\ell-1} = D_0 + H_{\ell-1}(\mathrm{ad}\,\eta \otimes T^*M).$$

(Henceforth, denote $H_{\ell-1}(\mathrm{ad}\,\eta \otimes T^*M)$ by $\Omega^1(\mathrm{ad}\,\eta)_{\ell-1}$.) We use $\ell - 1$ because of the differentiation in formula (2.1). It follows from this formula that $\mathcal{G}_\ell$ acts smoothly on $\mathcal{A}_{\ell-1}$ (cf. $(A.3)$).

3. The curvature operator

$$F : \mathcal{A}_{\ell-1} \to \Omega^2(\mathrm{ad}\,\eta)_{\ell-2}$$

is smooth (cf. $(A.4)$). The differential of F, written δF, at $D \in \mathcal{A}_{\ell-1}$ is the linear map

$$D : \Omega^1(\mathrm{ad}\,\eta)_{\ell-1} \to \Omega^2(\mathrm{ad}\,\eta)_{\ell-2}.$$

4. If F_D is self-dual, then it follows from elliptic regularity arguments which we outline in §8 that there is an element $s \in \mathcal{G}_\ell$ for which $s^*(D)$ is a smooth connection (cf. [**Pa**, §5]). This property ensures that the topology of $\hat{\mathcal{M}} \subset \hat{\mathcal{A}}_{\ell-1}/\mathcal{G}_\ell$ is independent of $\ell > 2$.

Reducible Connections.

We say that $D \in \mathcal{A}_{\ell-1}$ is *reducible* (or *split*) if the bundle $\eta = \lambda_1 \oplus \lambda_2$ and the connection $D = d_1 \oplus d_2$ both decompose. Such splittings correspond to singularities in the moduli space. It is important, then, to know when D is reducible.

THEOREM 3.1. *Assume that D is not flat $(F_D \not\equiv 0)$. Then the following are equivalent:*

    (a) $\mathcal{G}_{\ell,D}/\mathbf{Z}_2 \simeq U(1)$, *where $\mathcal{G}_{\ell,D} \subseteq \mathcal{G}_\ell$ is the stabilizer of D;*
    (b) $D : \Omega^0(\mathrm{ad}\,\eta)_\ell \to \Omega^1(\mathrm{ad}\,\eta)_{\ell-1}$ *has a nonzero kernel;*
    (c) D *is reducible;*
    (d) $\mathcal{G}_{\ell,D}/\mathbf{Z}_2 \neq 1$.

The $\mathbf{Z}_2$ appearing in the theorem is the center of $\mathcal{G}_\ell$ (cf. §2).

PROOF: (a) $\Rightarrow$ (b). Choose a nonzero element $u \in \Omega^0(\mathrm{ad}\,\eta)_\ell$ in the Lie algebra of $\mathcal{G}_{\ell,D}$. Differentiating the trivial action of the gauge transformations $\mathcal{G}_{\ell,D}$ on D, we obtain $u \circ D = D \circ u$, or $Du = 0$.

(b) $\Rightarrow$ (c). Fix $u \in \mathrm{Ker}\,D$. Then as u is pointwise traceless skew-Hermitian, its eigenvalues are $\pm i\lambda$. In an open set where $\lambda > 0$, choose a smoothly varying eigenvector e with $ue = i\lambda e$ and $(e, e) = 1$. Differentiating these equations, we obtain

$$uDe = i(d\lambda)e + i\lambda De,$$

$$Re(De, e) = 0.$$

Take the inner product of the first equation with e, and then examine the imaginary part. Thus

$$d\lambda = \mathrm{Im}(uDe, e) = -\,\mathrm{Im}(De, ue) = \lambda Re(De, e) = 0.$$

It follows that λ is constant and e is globally defined. This gives a splitting $\eta = \lambda_1 \oplus \lambda_2$. Furthermore, the equations above now imply that $De = 0$, so that $D = d_1 \oplus d_2$ is a reducible connection.

(c) $\Rightarrow$ (d). The circle $\begin{pmatrix} e^{i\theta} & 0 \\ 0 & e^{-i\theta} \end{pmatrix}$ is contained in $\mathcal{G}_{\ell,D}$. More precisely, the principal $SU(2)$ bundle P associated to η reduces to a $U(1)$ bundle Q. Since $U(1)$ is abelian, the adjoint bundle $Q \times_{U(1)} U(1)$ is trivial, has the trivial connection (induced by the connection on Q), and is included in $P \times_{SU(2)} SU(2)$. The circle of constant sections of $Q \times_{U(1)} U(1)$ is contained in $\mathcal{G}_{\ell,D}$.

(d) $\Rightarrow$ (a). Let s be a section of $P \times_{SU(2)} SU(2)$ with $Ds = 0$, i.e., $s \in \mathcal{G}_{\ell,D}$. If $s \neq \pm id$, then s has unequal constant eigenvalues as above, so that the eigenspaces define a splitting $\eta = \lambda_1 \oplus \lambda_2$. As before, we see that s lies in a circle action of $\mathcal{G}_{\ell,D}$. If $\mathcal{G}_{\ell,D}/\mathbb{Z}_2$ is larger than $U(1)$, then the holonomy group of D, which is centralized by the stabilizer $\mathcal{G}_{\ell,D}$, is smaller than $U(1)$, hence discrete. But then $F_D \equiv 0$, which contradicts our hypothesis.

As a corollary of the proof we see that $\mathcal{G}_\ell/\mathbb{Z}_2$ acts freely on $\hat{\mathcal{A}}$. Also, if D is reducible, then $\text{Ker}\, D$ is one dimensional. We repeatedly use the characterization (b) of reducible connections in the succeeding arguments.

A Slice Theorem.

The following are basic facts about group actions [V, §2.9]. Suppose G is a group acting on a manifold M, $\Gamma = \{\langle x, g \cdot x \rangle : x \in M, g \in G\} \subseteq M \times M$ is the *graph* of the action, and M/G is the orbit space with the quotient topology. Then

(i) M/G is Hausdorff if and only if Γ is closed.

(ii) If through each $x \in M$ we can find a *slice* of the action—that is, an open submanifold N containing x such that (a) $T_y M = T_y(G \cdot y) \oplus T_y N$ for all $y \in N$, (b) the restriction of the projection $M \to M/G$ to N is one-to-one—and if the action of G is free, then M/G is a manifold (which is Hausdorff if (i) is satisfied).

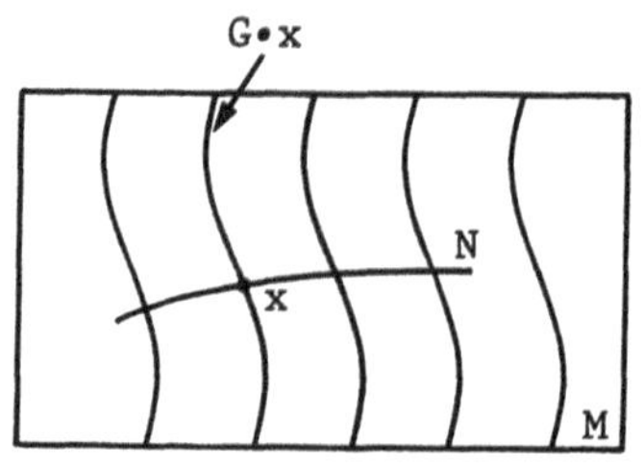

Consider the orbit space $\hat{\mathcal{X}}_{\ell-1}$ arising from the action of gauge transformations $\mathcal{G}_\ell$ on irreducible connections $\hat{\mathcal{A}}_{\ell-1}$. We may as well replace $\mathcal{G}_\ell$ by $\tilde{\mathcal{G}}_\ell = \mathcal{G}_\ell/\mathbf{Z}_2$, since $\tilde{\mathcal{G}}_\ell$ acts freely on $\tilde{\mathcal{A}}_{\ell-1}$. We drop the Sobolev subscripts momentarily to motivate our construction. Fix $D \in \hat{\mathcal{A}}$. Since $\mathcal{A}$ has an inner product, the natural slice to try is the orthogonal complement to the tangent along the orbit $\mathcal{G} \cdot D$. Taking D as base we can write an arbitrary connection as $D + A$ with $A \in \Omega^1(\mathrm{ad}\,\eta)$. It suffices to consider the Lie algebra $\Omega^0(\mathrm{ad}\,\eta)$ of $\tilde{\mathcal{G}}$. The action by $u \in \Omega^0(\mathrm{ad}\,\eta)$ is $D \mapsto Du$. So we seek the set of connections A which satisfy

$$(A, Du) = 0,$$

or

$$(D^*A, u) = 0,$$

for all u. Therefore,

$$\mathcal{X}_D = \{A \in \Omega^1(\mathrm{ad}\,\eta)_{\ell-1} : D^*A = 0\}$$

is our slice (translated to the origin), which we think of as the tangent space to $\hat{\mathcal{X}}$ at D.

THEOREM 3.2. *At each $D \in \hat{\mathcal{A}}_{\ell-1}$, the space $\hat{\mathcal{A}}_{\ell-1}$ is locally diffeomorphic to $\mathrm{Ker}(D^*) \times \tilde{\mathcal{G}}_\ell$. In other words, there exists a neighborhood $\mathcal{O}_D$ of D in $\hat{\mathcal{A}}_{\ell-1}$ and a map $s : \mathcal{O}_D \to \tilde{\mathcal{G}}_\ell$ such that $s(D')^*D' \in D + \mathcal{X}_D$ for $D' \in \mathcal{O}_D$. Moreover,*

$$\Phi_D : D' \mapsto \langle s(D')^*D' - D, s(D')\rangle$$

maps $\mathcal{O}_D$ diffeomorphically onto a neighborhood of $\langle 0, id\rangle$ in $\mathcal{X}_D \times \tilde{\mathcal{G}}_\ell$. Finally, Φ_D is equivariant with respect to the action of $\tilde{\mathcal{G}}_\ell$ on $\hat{\mathcal{A}}_{\ell-1}$, and $\mathcal{O}_D$ can be chosen to be $\tilde{\mathcal{G}}_\ell$ invariant.

PROOF: As before, write $D' = D + A$. The map s is constructed by solving the equation

$$L_D(A, s) = D^*(s^{-1}Ds + s^{-1}As) = 0.$$

The differential of L_D at $\langle 0, id\rangle$ is given by

$$\delta L_D : \Omega^1(\mathrm{ad}\,\eta)_{\ell-1} \oplus \Omega^0(\mathrm{ad}\,\eta)_\ell \to \Omega^0(\mathrm{ad}\,\eta)_{\ell-2}$$
$$\delta A \oplus \delta s \mapsto D^*(D(\delta s) + \delta A).$$

The partial differential $\delta_2 L_D = D^*D$ in the second factor is a self-adjoint elliptic operator. Since D is irreducible, the results of the previous section show D^*D has no kernel, whence by standard elliptic theory D^*D is invertible. From the implicit function theorem in Hilbert space we obtain the neighborhood $\mathcal{O}_D$ and the map s. The inverse of Φ_D is given explicitly as

$$\langle A', s \rangle \mapsto s A' s^{-1} - (Ds)s^{-1}.$$

Equivariance follows from the equation

$$L_D(t^* A, t^{-1} s) = L_D(A, s), \qquad t \in \mathcal{G}.$$

COROLLARY. $\hat{\mathcal{X}}_{\ell-1}$ is a Hausdorff manifold, if $\ell > 2$.

PROOF: We have to show that $\hat{\mathcal{X}}_{\ell-1}$ is Hausdorff and that the restriction of the projection $\hat{\mathcal{A}}_{\ell-1} \to \hat{\mathcal{A}}_{\ell-1}/\tilde{\mathcal{G}}_\ell$ to a neighborhood of the origin in $D + \mathcal{X}_D$ is one-to-one. We first prove the Hausdorff property by showing that

$$\Gamma = \{\langle D, s^*(D)\rangle : D \in \hat{\mathcal{A}}_{\ell-1}, s \in \mathcal{G}_\ell\} \subseteq \hat{\mathcal{A}}_{\ell-1} \times \hat{\mathcal{A}}_{\ell-1}$$

is closed. Suppose the sequence $\{\langle (D+A_n), s_n^*(D+A_n)\rangle\} \subseteq \Gamma$ is convergent. Then

$$s_n^*(D + A_n) = D + s_n^{-1}(Ds_n) + s_n^{-1} A_n s_n$$
$$= D + A_n',$$

so that $A_n' \to A'$ for some $A' \in \Omega^1(\mathrm{ad}\,\eta)_{\ell-1}$. It follows from this and from the fact that $A_n \to A$ for some $A \in \Omega^1(\mathrm{ad}\,\eta)_{\ell-1}$ that $\{A_n\}$ and $\{A_n'\}$ are bounded in $\Omega^1(\mathrm{ad}\,\eta)_{\ell-1}$. Now the inclusion $\mathrm{Aut}\,\eta \subseteq \mathrm{End}\,\eta$ (i.e., $SU(2) \subseteq \mathfrak{gl}(2;\mathbb{C})$) gives an embedding $\mathcal{G}_\ell \hookrightarrow H_\ell(\mathrm{End}\,\eta)$. Also, $\mathrm{ad}\,\eta \hookrightarrow \mathrm{End}\,\eta$ and the inner product

$$(X, Y) = tr(XY^*)$$

on $\mathfrak{gl}(2;\mathbb{C})$ extends (2.12), so that this embedding is an isometry. We view s_n, A_n, A_n' as sections of $\mathrm{End}\,\eta$ so that we can estimate the size of s_n with Sobolev norms. From

$$(3.3) \qquad\qquad Ds_n = s_n A_n' - A_n s_n$$

and the fact that s_n is unitary, we see first that $\|Ds_n\|_0$ is bounded, and by bootstrapping with (3.3) that $\|Ds_n\|_1, \|Ds_n\|_2, \ldots, \|Ds_n\|_{\ell-1}$ are bounded.

$(\|_k$ is the norm in $H_k(\operatorname{End}\eta \otimes T^*M)$.) Details can be found in Proposition A.5 of Appendix A. Hence $\{s_n\}$ is bounded in $H_\ell(\operatorname{End}\eta)$, so by Rellich's Theorem there is a subsequence (call it $\{s_n\}$ again) covergent in $H_{\ell-1}(\operatorname{End}\eta)$. Now (3.3) shows that $\{Ds_n\}$ converges in $H_{\ell-1}(\operatorname{End}\eta \otimes T^*M)$ so that $\{s_n\}$ converges to $s \in H_\ell(\operatorname{End}\eta)$. Since $\mathcal{G}_\ell$ is closed in $H_\ell(\operatorname{End}\eta)$, $s \in \mathcal{G}_\ell$. By continuity, $s^*(D + A) = D + A'$, and Γ is closed.

Next we prove that for ε sufficiently small, if $A, A' \in \mathcal{X}_D$ satisfy $\|A\|_{\ell-1}$, $\|A'\|_{\ell-1} < \varepsilon$ and $D+A' = s^*(D+A)$ for some $s \in \mathcal{G}_\ell$, then $\|s\pm 1\|_\ell$ is so small (for one choice of sign) that $\langle A', s\rangle \in \Phi_D(\mathcal{O}_D)$. It then follows from the theorem that $A = A'$, whence the ball of radius ε in $D + \mathcal{X}_D$ injects into the quotient (i.e., is a coordinate chart on $\hat{\mathcal{A}}_{\ell-1}/\tilde{\mathcal{G}}_\ell$). First, decompose $s = s_0 + c$ in $H_\ell(\operatorname{End}\eta)$ into components $c \in \operatorname{Ker} D$ and $s_0 \in (\operatorname{Ker} D)^\perp$. Note that c is a constant diagonal transformation, and that $s_0, c \notin \mathcal{G}_\ell \subset H_\ell(\operatorname{End}\eta)$ in general. Estimate s_0 by bootstrapping as in the previous argument (cf. (A.5)), only now keep track of the accumulated constants. The result is $\|s_0\|_\ell < K\varepsilon$ for some K depending on D and ℓ. Since $|s| = 1$ (s is unitary) and $|s_0|$ is small, it follows that $|c| \simeq 1$. Replace c by $c/|c| = \pm 1$ and adjust s_0 accordingly. Then the previous inequality continues to hold with $2K$ replacing K. This gives the desired inequality $\|s \pm 1\|_\ell < 2K\varepsilon$.

The Parametrized Moduli Space.

Our ultimate goal in this chapter is to prove that the moduli space is a manifold. Unfortunately, this may not be true; we may have to perturb the metric. We could apply transversality directly were it not for the gauge transformations. Our argument essentially amounts to an equivariant transversality theorem. Consider, then, the union of moduli spaces over all metrics. We encounter technical difficulities if we parametrize the space of metrics directly; we lose control of self-duality. Rather, we choose as a parameter space the set $\mathcal{C} = C^k(GL(TM))$ of C^k automorphisms of the tangent bundle. Of course, this is just the gauge group for the bundle of frames. We use C^k since then $\mathcal{C}$ is a Banach manifold. (It is only a Fréchet manifold if we use C^∞.) Now $\mathcal{C}$ acts on all covariant tensors of order two (sections of $T^*M \otimes T^*M$) by pullback. So if g is our base metric on M, and $\varphi \in \mathcal{C}$, then φ^*g is a new metric, and every metric is realized in this fashion, uniquely if we require that φ be symmetric. This follows from the realization of inner products on $\mathbf{R}^n$ as the homogeneous space $GL(n)/O(n)$. Of course, since we allow all φ, we realize each metric many times. This does not affect genericity arguments at all. Let

$$P_- : \Omega^2 \to \Omega^2_-$$

$$\theta \mapsto \frac{1}{2}(\theta - *\theta)$$

denote the projection onto anti-self-dual 2-forms with respect to g; then $\varphi^* P_-(\varphi^{-1})^*$ is the projection onto anti-self-dual 2-forms with respect to $\varphi^* g$. Fix $k >> \ell$ and define

$$\mathcal{P} : \hat{\mathcal{A}}_{\ell-1} \times \mathcal{C} \to \Omega^2_-(\operatorname{ad}\eta)_{\ell-2}$$

$$\langle D, \varphi \rangle \mapsto P_-((\varphi^{-1})^* F_D).$$

Then $\mathcal{P}(D,\varphi) = 0$ if and only if F_D is self-dual with respect to $\varphi^* g$. ($\mathcal{C}$ was chosen as our parameter space precisely so that we could detect self-duality by mapping into a fixed space.)

THEOREM 3.4. *The map $\mathcal{P}$ is smooth and has zero as a regular value.*

This means that the differential $\delta\mathcal{P}_{\langle D,\varphi\rangle}$ is onto when $\mathcal{P}(D,\varphi) = 0$. It follows from this theorem that $\mathcal{P}^{-1}(0) = \mathcal{S}\hat{\mathcal{D}}_{\ell-1}$ is a smooth manifold. This is not quite our parametrized moduli space; we still have to divide out the action of gauge transformations.

As a preliminary step toward proving Theorem 3.4, we investigate how the infinitesimal variations of frame act on self-dual and anti-self-dual 2-forms. Let V be a 4-dimensional oriented Euclidean vector space, thought of as the tangent space to M at a point, and fix an oriented orthonormal basis $\{e^k\}$ for V^*. let σ^{ij} be the orthogonal projection of $e^i \wedge e^j$ into the self-dual forms $\Lambda^2_+ V^*$, i.e., $\sigma^{ij} = P_+(e^i \wedge e^j)$, and similarly let α^{ij} be the orthogonal projection of $e^i \wedge e^j$ into the anti-self-dual forms $\Lambda^2_- V^*$, $\alpha^{ij} = P_-(e^i \wedge e^j)$. Then a basis for $\Lambda^2_+ V^*$ is $\{\sigma^{12}, \sigma^{13}, \sigma^{14}\} = \{\sigma^{34}, \sigma^{42}, \sigma^{23}\}$, and a basis for $\Lambda^2_- V^*$ is $\{\alpha^{12}, \alpha^{13}, \alpha^{14}\} = \{\alpha^{43}, \alpha^{24}, \alpha^{32}\}$. The following simple lemma shows that we can rotate coordinates so that an arbitrary (anti-) self-dual form is a basis element.

LEMMA 3.5. *If $\theta_\pm \in \Lambda^2_\pm V^*$, then there exists an orthonormal basis $\{e^k\}$ of V^* such that*

$$\theta_\pm = \frac{|\theta_\pm|}{\sqrt{2}}(e^1 \wedge e^2 \pm e^3 \wedge e^4).$$

PROOF: It suffices to prove that the orbits of the $SO(4)$ action on $\Lambda^2_\pm V^*$ are spheres. This can be seen by group theory—under the global decomposition $\mathrm{Spin}(4) = \mathrm{Spin}(3) \times \mathrm{Spin}(3)$ corresponding to $\mathfrak{so}(4) = \mathfrak{so}(3) \oplus \mathfrak{so}(3)$, $\Lambda^2_\pm V^*$ is the adjoint representation of a $\mathrm{Spin}(3)$ factor, i.e., the standard representation of $SO(3)$—or by fixing e^1 arbitrarily, and then explicitly rotating the coordinates to the desired form.

Infinitesimal changes in metric are effectively generated by symmetric changes in frame, i.e., by transformations in $\mathfrak{gl}(V)$ which are symmetric with respect to the fixed metric on V. Clearly, the skew-symmetric matrices are infinitesimal rotations and do not affect the metric. The symmetric transformations are spanned by

$$r^i_i = e_i \otimes e^i,$$

$$r^i_j = e_i \otimes e^j + e_j \otimes e^i, \qquad i \neq j,$$

and act on $\Lambda^2 V^*$ as derivations extending the transpose $r^* \in \mathrm{End}(V^*)$. Explicitly,

$$(r^i_i)^*(e^i \wedge e^j) = e^i \wedge e^j,$$

$$(r^i_i)^*(e^j \wedge e^k) = 0,$$

$$(r^i_k)^*(e^k \wedge e^j) = e^i \wedge e^j,$$

$$(r^i_k)^*(e^i \wedge e^k) = 0,$$

$$(r^i_k)^*(e^\ell \wedge e^j) = 0,$$

where i, j, k, ℓ are unequal indices in each equation. Then

$$(3.6) \qquad \begin{aligned} (r^i_i)^* \sigma^{ij} &= \sigma^{ij} + \alpha^{ij}, \\ (r^i_k)^* \sigma^{ij} &= \alpha^{kj}, \qquad i \neq j \neq k, \end{aligned}$$

and formulas for anti-self-dual forms follow by reversing the orientation. The action extends trivially to $\Lambda^2_\pm V^* \otimes W$ for any vector space W; in our application, W is a fiber of the adjoint bundle. For the next lemma, assume that W is endowed with an inner product.

LEMMA 3.7. *Suppose $F \in \Lambda^2_+ V^* \otimes W$ and $\Phi \in \Lambda^2_- V^* \otimes W$ satisfy $(r^*F, \Phi) = 0$ for all $r \in \mathfrak{gl}(V)$. Then $(F, \Phi)_W \in \Lambda^2_+ V^* \otimes \Lambda^2_- V^*$ is equal to zero.*

The conclusion reads better in coordinates. Write $F = \sum_{i=2}^4 \sigma^{1i} \otimes F_i$ and $\Phi = \sum_{j=2}^4 \alpha^{1j} \otimes \Phi_j$; then we conclude $(F_i \Phi_j) = 0$ for all i, j. Yet again, if we view $F \in \mathrm{Hom}(\Lambda^2_+ V, W)$ and $\Phi \in \mathrm{Hom}(\Lambda^2_- V, W)$, then the lemma says that the images $\mathrm{Im}(F)$ and $\mathrm{Im}(\Phi)$ are orthogonal.

PROOF: For $r = r^1_3 + r^2_4$,

$$(r^1_3 + r^2_4)^* \sigma^{12} = 2\alpha^{14},$$
$$(3.8) \qquad (r^1_3 + r^2_4)^* \sigma^{13} = 0,$$
$$(r^1_3 + r^2_4)^* \sigma^{14} = 0.$$

Thus,

$$(3.9) \qquad 0 = ((r^1_3 + r^2_4)^* F, \Phi) = 2(F_2, \Phi_4).$$

By symmetry, $(F_2, \Phi_3) = 0$. Now for $r = r^1_1 + r^2_2 - r^3_3 - r^4_4$,

$$(r^1_1 + r^2_2 - r^3_3 - r^4_4)^* \sigma^{12} = 2\alpha^{12},$$
$$(3.10) \qquad (r^1_1 + r^2_2 - r^3_3 - r^4_4)^* \sigma^{13} = 0,$$
$$(r^1_1 + r^2_2 - r^3_3 - r^4_4)^* \sigma^{14} = 0.$$

Consequently,

$$(3.11) \qquad 0 = ((r^1_1 + r^2_2 + r^3_3 - r^4_4)^* F, \Phi) = 2(F_2, \Phi_2).$$

Collecting (3.9) and (3.11), $(F_2, \Phi_j) = 0$ for all j, and by symmetry all (F_i, Φ_j) vanish.

Our next lemma is an elementary property of $\mathfrak{su}(2)$.

LEMMA 3.12. *For nonzero* $X, Y \in \mathfrak{su}(2)$, *if* $(X, Y) = 0$ *then* $[X, Y] \neq 0$.

PROOF: With respect to the basis

$$\sigma_1 = \begin{pmatrix} i & 0 \\ 0 & -i \end{pmatrix}, \quad \sigma_2 = \begin{pmatrix} 0 & 1 \\ -1 & 0 \end{pmatrix}, \quad \sigma_3 = \begin{pmatrix} 0 & i \\ i & 0 \end{pmatrix}$$

of $\mathfrak{su}(2)$, write $X = X^i \sigma_i$ and $Y = Y^i \sigma_i$. Then $(X, Y) = -tr(XY) = 2X \cdot Y$, and $[X, Y] = 2X \times Y$ where $\cdot$ and $\times$ are the dot and cross products of vectors in $\mathbf{R}^3$. If $X \cdot Y = X \times Y = 0$, then $X = 0$ or $Y = 0$.

PROOF OF THEOREM 3.4: $\mathcal{P}$ is quadratic in ΛD and φ, and hence is easily seen to be smooth. Let $\mathfrak{c} = C^k(\mathrm{End}(TM))$ denote the Lie algebra of

C. We split the differential of $\mathcal{P}$ into its two factors along connections and frames:

$$\begin{array}{ccccc}
\delta\mathcal{P}_{\langle D,\varphi\rangle} & : & T_D(\mathcal{A}_{\ell-1}) & \oplus\ T_\varphi(\mathcal{C}) & \to \Omega^2_-(\mathrm{ad}\,\eta)_{\ell-2} \\
\| & & \| & \| & \| \\
\delta_1\mathcal{P}_{\langle D,\varphi\rangle} \oplus \delta_2\mathcal{P}_{\langle D,\varphi\rangle} & : & \Omega^1(\mathrm{ad}\,\eta)_{\ell-1} \oplus & \mathfrak{c} & \to \Omega^2_-(\mathrm{ad}\,\eta)_{\ell-2}.
\end{array}$$

Then

$$\delta_1\mathcal{P} = \delta_1\mathcal{P}_{\langle D,\varphi\rangle} : \Omega^1(\mathrm{ad}\,\eta)_{\ell-1} \to \Omega^2_-(\mathrm{ad}\,\eta)_{\ell-2}$$

$$A \mapsto P_-((\varphi^{-1})^* DA)$$

and

$$\delta_2\mathcal{P} = \delta_2\mathcal{P}_{\langle D,\varphi\rangle} : \mathfrak{c} \to \Omega^2_-(\mathrm{ad}\,\eta)_{\ell-2}$$

$$r \mapsto P_-((\varphi^{-1})^*(r^* F_D))$$

are the partial differentials of $\mathcal{P}$. (Note that $(\varphi^{-1})^*$ is a Lie group action and r^* is a Lie algebra action.)

Suppose that $\mathcal{P}(D,\varphi) = 0$. We want to show that $\delta\mathcal{P}$ is onto, or equivalently $\mathrm{Coker}(\delta\mathcal{P}) = 0$. We identify $\mathrm{Coker}(\delta\mathcal{P})$ with a subspace of $\Omega^2_-(\mathrm{ad}\,\eta)_{2-\ell}$, the L^2 dual of $\Omega^2_-(\mathrm{ad}\,\eta)_{\ell-2}$. Now $\delta_1\mathcal{P}$ is surjective elliptic (i.e., its symbol is onto) since it fits into the elliptic complex

$$0 \to \Omega^0(\mathrm{ad}\,\eta)_\ell \xrightarrow{D} \Omega^1(\mathrm{ad}\,\eta)_{\ell-1} \xrightarrow{\delta_1\mathcal{P}} \Omega^2_-(\mathrm{ad}\,\eta)_{\ell-2} \to 0.$$

So $\mathrm{Im}(\delta_1\mathcal{P})$ is closed and has finite codimension. (The metrics for which $\delta_1\mathcal{P}$ is surjective are precisely those for which $\hat{\mathcal{M}}$ is a manifold.) Then $\mathrm{Im}(\delta\mathcal{P}) \supseteq \mathrm{Im}(\delta_1\mathcal{P})$ is also closed of finite codimension. Furthermore, $\mathrm{Coker}(\delta\mathcal{P}) \subseteq \mathrm{Coker}(\delta_1\mathcal{P})$ consists of functions in $\Omega^2_-(\mathrm{ad}\,\eta)_{\ell-2}$, rather than just distributions in $\Omega^2_-(\mathrm{ad}\,\eta)_{2-\ell}$. This follows from regularity for the overdetermined elliptic system $(\delta_1\mathcal{P})^*\Phi = 0$ (which has C^k coefficients).

Assume that $\Phi \in \mathrm{Coker}(\delta\mathcal{P})$; we will show $\Phi = 0$. Then $\Phi \in \mathrm{Coker}(\delta_1\mathcal{P})$ so that

$$0 = \int_M (P_-(\varphi^{-1})^* DA, \Phi)_g$$

$$= \int_M (DA, \varphi^*(\Phi))_{\varphi^* g}$$

$$= \int_M (A, D^*\tilde{\Phi})_{\varphi^* g}$$

for all $A \in \Omega^1(\operatorname{ad}\eta)_{\ell-1}$. Here $\tilde{\Phi} = \varphi^*(\Phi)$ and D^* is the adjoint of D in the metric $\varphi^* g$. Since Φ is a continuous function (from the elliptic regularity above), we have the pointwise equation

$$(3.13) \qquad\qquad D^*\tilde{\Phi} = 0.$$

Also, $\Phi \in \operatorname{Coker}(\delta_2 \mathcal{P})$ so that

$$0 = \int_M (P_-((\varphi^{-1})^*(r^* F_D)), \Phi)_g$$
$$= \int_M (r^* F_D, \tilde{\Phi})_{\varphi^* g}$$

for all $r \in \mathfrak{c}$. Again we can conclude that

$$(3.14) \qquad\qquad (r^* F_D, \tilde{\Phi})_{\varphi^* g} = 0$$

pointwise. It follows from (3.14) and Lemma 3.7 that $\operatorname{Im}(F_D)$ and $\operatorname{Im}(\tilde{\Phi})$ are perpendicular (pointwise). In particular, at each point where F_D and $\tilde{\Phi}$ are nonzero, one of F_D and $\tilde{\Phi}$ has rank 1 as an endomorphism.

Henceforth, we use the metric $\varphi^* g$; then $F = F_D$ is self-dual, $\tilde{\Phi}$ is anti-self-dual, and $DF = D^*F = D\tilde{\Phi} = D^*\tilde{\Phi} = 0$. Hence both F and $\tilde{\Phi}$ are annihilated by $DD^* + D^*D$. Unique continuation for linear elliptic differential operators [**Ar**], extended to elliptic systems with scalar symbol, implies that neither F nor $\tilde{\Phi}$ vanishes on an open set unless it vanishes identically. Suppose now that $\tilde{\Phi} \not\equiv 0$ and that at some point where $\tilde{\Phi} \neq 0$, F has rank 2. Then in some neighborhood of that point F has rank 2 and $\tilde{\Phi}$ has rank 1. Write $\tilde{\Phi} = \alpha \otimes u$ for $\alpha(x) \in \Omega^2_-(x)$ and $u(x) \in \operatorname{ad}\eta_x$ with $|u(x)| = 1$. Now

$$(3.15) \qquad\qquad 0 = D\tilde{\Phi} = d\alpha \otimes u + \alpha \wedge Du.$$

From $|u| = 1$, we have

$$0 = \frac{1}{2} d(u, u) = (Du, u).$$

Therefore,

$$d\alpha = -\alpha \wedge (Du, u) = 0,$$

and so

$$\alpha \wedge Du = 0.$$

By Lemma 3.5 we can choose a coframe θ^i (at a point) so that

$$\alpha = \frac{|\alpha|}{\sqrt{2}}(\theta^1 \wedge \theta^2 - \theta^3 \wedge \theta^4).$$

Then if

$$Du = \sum_i \theta^i \otimes w_i, \qquad w_i \in \mathrm{ad}\,\eta,$$

we have

$$0 = \alpha \wedge Du = \frac{|\alpha|}{\sqrt{2}}(\theta^1 \wedge \theta^2 \wedge \theta^3 \otimes w_3 + \theta^1 \wedge \theta^2 \wedge \theta^4 \otimes w_4$$
$$- \theta^3 \wedge \theta^4 \wedge \theta^1 \otimes w_1 - \theta^3 \wedge \theta^4 \wedge \theta^2 \otimes w_2),$$

so that $w_1 = w_2 = w_3 = w_4 = 0$, whence $Du = 0$. (This argument—that if $D(\alpha \otimes u) = 0$ and $|u| = 1$, then $Du = 0$—will be used again in §4, only there α will be replaced by a self-dual 2-form. Another fact gleaned here is that the wedge product $\alpha : \Omega^1(\mathrm{ad}\,\eta) \to \Omega^3(\mathrm{ad}\,\eta)$ is invertible.)

Now $D^2u = [F, u] \neq 0$ by Lemma 3.12 since $(F, u) = 0$. This contradiction (to $Du = 0$) shows that F has rank at most 1. On the dense open set where $F \neq 0$ we write $F = \sigma \otimes u$ with $|u| = 1$; then $Du = 0$ as before. If the complement of $\{F = 0\}$ is connected, then we can consistently extend u over $\{F = 0\}$ as $Du = 0$. Since the connection can acquire no holonomy on the set $F = 0$, this definition is consistent on the closure of a connected set. But this contradicts (b) of Theorem 3.1 since D is not reducible. All we have left to show, then, is that the complement of $\{F = 0\}$ is connected. Suppose not. Then on a component U, the curvature F is in the kernel of the self-adjoint operator $DD^* + D^*D$ and vanishes identically on the boundary ∂U. Therefore, by the variational characterization of eigenvalues of self-adjoint operators, the first eigenvalue of $DD^* + D^*D$ on the larger domain M is negative [**CH**, p. 409]. (This is a musical phenomenon: the fundamental pitch of a trombone drops as the slide is extended.) But $DD^* + D^*D$ is a positive operator, a contradiction.

There is an alternative argument for the last step which proves a stronger assertion: The set $\{F = 0\}$ does not disconnect any domain $\mathcal{O} \subseteq M$. Otherwise we can

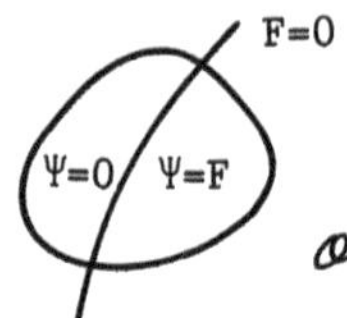

define a 2-form Ψ in $\mathcal{O}$ which equals F on one component and equals zero on the other. Since F is smooth, we can apply Stokes' Theorem to show that Ψ has a distributional derivative, and therefore $\Psi \in H_1(\mathcal{O})$. It follows that Ψ satisfies the *first order* equations $D\Psi = D^*\Psi = 0$, and by elliptic regularity for this system, Ψ is smooth. But Ψ vanishes on an open set, violating the Unique Continuation Principle. This contradiction proves that F does not vanish on codimension one submanifolds (just as holomorphic functions do not vanish on curves).

We must now reduce the large manifold of self-dual connections $\mathcal{S}\hat{\mathcal{D}}_{\ell-1}$ to a finite dimensional moduli space. The first stage is easy.

THEOREM 3.16. $\mathcal{S}\hat{\mathcal{D}}_{\ell-1}/\mathcal{G}_\ell \subseteq \hat{\mathcal{X}}_{\ell-1} \times \mathcal{C}$ *is a manifold.*

PROOF: Fix $\langle D, \varphi \rangle \in \mathcal{S}\hat{\mathcal{D}}_{\ell-1} = \mathcal{P}^{-1}(0)$. As in Theorem 3.1 we construct a slice of the action by restricting our attention to connections annihilated by D^*. More precisely, restrict $\mathcal{P}$ to

$$\overline{\mathcal{P}} : \{D + A \in \mathcal{A}_{\ell-1} : D^*A = 0\} \times \mathcal{C} \to \Omega^2_-(\operatorname{ad}\eta)_{\ell-2},$$

where the adjoint at $\langle D + A, \varphi \rangle$ refers to the metric φ^*g. Consider the elliptic complex

$$0 \to \Omega^0(\operatorname{ad}\eta)_\ell \xrightarrow{D} \Omega^1(\operatorname{ad}\eta)_{\ell-1} \xrightarrow{\delta_1 \mathcal{P}} \Omega^2_-(\operatorname{ad}\eta)_{\ell-2} \to 0.$$

Now $\operatorname{Ker} D^* \perp \operatorname{Im} D$ and $\delta_1\mathcal{P}_{\langle D,\varphi\rangle}|_{\operatorname{Im} D} \equiv 0$. Since $\delta_1\overline{\mathcal{P}}_{\langle D,\varphi\rangle} = \delta_1\mathcal{P}_{\langle D,\varphi\rangle}|_{\operatorname{Ker} D^*}$, $\delta_2\overline{\mathcal{P}}_{\langle D,\varphi\rangle} = \delta_2\mathcal{P}_{\langle D,\varphi\rangle}$, and $\delta\mathcal{P}_{\langle D,\varphi\rangle}$ is onto, it follows that $\delta\overline{\mathcal{P}}_{\langle D,\varphi\rangle}$ is onto. Furthermore, surjectivity is an open condition, and so $\delta\overline{\mathcal{P}}_{\langle\cdot,\cdot\rangle}$ is onto in a neighborhood $\mathcal{O}$ of $\langle D, \varphi \rangle$. Therefore, $(\overline{\mathcal{P}})^{-1}(0) \cap \mathcal{O}$ is a manifold. By the construction in Theorem 3.2 we can identify this with a neighborhood of $\langle D, \varphi \rangle$ in the orbit space $\hat{\mathcal{X}}_{\ell-1}$. Thus $\mathcal{S}\hat{\mathcal{D}}_{\ell-1}/\mathcal{G}_\ell$ is a submanifold of $\hat{\mathcal{X}}_{\ell-1} \times \mathcal{C}$ (and therefore Hausdorff).

The quotient $\mathcal{S}\hat{\mathcal{D}}_{\ell-1}/\mathcal{G}_\ell$ is our parametrized moduli space.

The Moduli Space.

The following diagram summarizes the various manifolds of connections obtained thus far:

$$\mathcal{P}^{-1}(0) = \mathcal{S}\hat{\mathcal{D}}_{\ell-1} \subseteq \hat{\mathcal{A}}_{\ell-1} \times \mathcal{C} \overset{\mathcal{P}}{\to} \Omega^2_-(\operatorname{ad}\eta)_{\ell-2}$$

$$\begin{array}{ccc}
\downarrow & & \downarrow \\
\mathcal{S}\hat{\mathcal{D}}_{\ell-1}/\mathcal{G}_\ell & \subseteq & \hat{\mathcal{X}}_{\ell-1} \times \mathcal{C} \\
\downarrow\, \pi & & \downarrow\, \pi \\
\mathcal{C} & = & \mathcal{C}
\end{array}$$

Notice that $\pi^{-1}(\varphi) = \hat{\mathcal{M}}_\varphi$ is the moduli space of self-dual connections for the metric $\varphi^* g$. We will prove

THEOREM 3.17. *There exists a Baire set of $\varphi \in \mathcal{C}$ for which $\hat{\mathcal{M}}_\varphi$ is a 5-manifold.*

A *Baire set* is a countable intersection of open dense sets. It follows that the set of good metrics is dense; a separate argument will prove that it is also open. Of course, $\hat{\mathcal{M}}_\varphi$ is a manifold when φ is a regular value of π. This fancy differential topology is just the ordinary Implicit Function Theorem, only in an infinite dimensional setting. Our machinery is set up so that Theorem 3.17 follows directly from the

SARD–SMALE THEOREM [Sm]. *Let $\pi : E \to \mathcal{C}$ be a Fredholm map between paracompact Banach manifolds. Then the set of regular values of π is a Baire set in $\mathcal{C}$.*

The map π is *Fredholm* if at each point of E the differential $\delta\pi$ is Fredholm (i.e., finite dimensional kernel, finite dimensional cokernel, and closed range) as a linear map between Banach spaces. The *index* of π, denoted $\operatorname{ind}(\pi)$, is the index of the differential $\operatorname{ind}(\delta\pi) = \dim \operatorname{Ker}(\delta\pi) - \dim \operatorname{Coker}(\delta\pi)$ at any point of E. Since the index is a deformation invariant, $\operatorname{ind}(\delta\pi)$ is independent of the point chosen if E is connected. When φ is a regular value, $\pi^{-1}(\varphi)$ is a manifold of dimension $\operatorname{ind}(\pi)$. For our application $E = \mathcal{S}\hat{\mathcal{D}}_{\ell-1}/\mathcal{G}_\ell$ and $\mathcal{C} = C^k(GL(TM))$ are paracompact Banach manifolds by construction. To prove Theorem 3.17 we have only to show that π is Fredholm of index 5.

PROOF OF THEOREM 3.17: We calculate the index at $\overline{\langle D, id\rangle}$ for convenience. (The bar indicates the orbit under gauge transformations.) Now the tangent space to $\mathcal{S}\hat{\mathcal{D}}_{\ell-1}$ at $\langle D, id\rangle$ is given by

$$T_{\langle D, id\rangle}(\mathcal{S}\hat{\mathcal{D}}_{\ell-1}) = \{\langle A, r\rangle \in \Omega^1(\operatorname{ad}\eta)_{\ell-1} \oplus \mathfrak{c} : \delta_1\mathcal{P}(A) + \delta_2\mathcal{P}(r) = 0\}.$$

The tangent space to the quotient $\mathcal{S}\hat{\mathcal{D}}_{\ell-1}/\mathcal{G}_\ell$ can be represented as the tangent space to our slice:

$$T_{\overline{\langle D, id\rangle}}(\mathcal{S}\hat{\mathcal{D}}_{\ell-1}/\mathcal{G}_\ell) = \{\langle A, r\rangle : \delta_1 \mathcal{P}(A) + \delta_2 \mathcal{P}(r) = 0 \text{ and } D^* A = 0\}.$$

Clearly $\delta\bar{\pi}(A, r) = r$. So

$$\operatorname{Ker} \delta\bar{\pi} = \{\langle A, r\rangle : \delta_1 \mathcal{P}(A) = D^* A = r = 0\},$$

and

$$\operatorname{Im} \delta\bar{\pi} = (\delta_2 \mathcal{P})^{-1}(\operatorname{Im} \delta_1 \mathcal{P}|_{\{D^* A = 0\}}).$$

Consider yet again the elliptic complex

$$(3.18) \qquad 0 \to \Omega^0(\operatorname{ad}\eta)_\ell \xrightarrow{D} \Omega^1(\operatorname{ad}\eta)_{\ell-1} \xrightarrow{\delta_1 \mathcal{P}} \Omega^2_-(\operatorname{ad}\eta)_{\ell-2} \to 0,$$

where $\delta_1 \mathcal{P} = P_- D$. Since $\delta_1 \mathcal{P} \circ D = 0$, we can write

$$(3.19) \qquad\qquad \operatorname{Im} \delta\bar{\pi} = (\delta_2 \mathcal{P})^{-1}(\operatorname{Im} \delta_1 \mathcal{P}).$$

Now $\delta_1 \mathcal{P}$ has closed range and finite cokernel of dimension h^2, the dimension of the second cohomology of (3.18). Since $\delta\mathcal{P}$ is onto, it follows easily from (3.19) that $\operatorname{Im} \delta\bar{\pi}$ is also closed and of codimension h^2. Also, $\operatorname{Ker} \delta\bar{\pi}$ (projected onto the first factor) is exactly the first cohomology of (3.18) and has dimension h^1. Theorem 3.1 implies that h^0 vanishes. Finally, the Atiyah–Singer Index Theorem computes the index of (3.18) as

$$h^0 - h^1 + h^2 = -5.$$

(See [**AHS**] for details.) Altogether, then, we have proved that $\delta\bar{\pi}$ is a Fredholm map whose index is $h^1 - h^2 = 5$. Now the Sard–Smale Theorem applies to complete the proof of Theorem 3.17.

In our arguments we found it convenient to use C^k metrics, k large. However, we wish to point out that we can, in the end, assume that the metric is C^∞, or even real analytic. This follows from the fact that the "good" C^k metrics actually form an *open* dense set. To see this note that h^2 is an upper-semicontinuous integer-valued function on $\mathcal{X}_{\ell-1} \times \mathcal{C}$ (cf. (4.15)). This means, in particular, that h^2 vanishes on open sets. We prove in §4 that for the m singular points $\{p_1, p_2, \ldots, p_m\} \subset \mathcal{M}$, just as for the irreducible connections in $\hat{\mathcal{M}}$, the vanishing of h^2 signals the existence of local coordinate charts, although the "chart" at p_i is a cone on $\mathbf{CP}^2$, not $\mathbf{R}^5$. Furthermore, we show that $h^2(p_i)$ vanishes on a dense open set of C^k metrics. In §8 we prove that for a certain subset $\mathcal{M}_\lambda \subset \mathcal{M}$, the difference $\mathcal{M}\backslash\mathcal{M}_\lambda$ is compact, and the results in §9 imply that h^2 vanishes on $\mathcal{M}_\lambda$. Furthermore, the estimates leading to these results are uniform over small changes in the C^k metric. Fix a good metric. Then we can cover a neighborhood U of $\mathcal{M}\backslash\mathcal{M}_\lambda$ in $\mathcal{X}_{\ell-1}$ by a finite number of open sets such that h^2 vanishes on U for all nearby metrics. Putting all of this together, using the fact that finite intersections of open dense sets are open and dense, and noting that the moduli space depends smoothly on the metric, we have

PROPOSITION 3.20. *The set of C^k metrics for which $\hat{\mathcal{M}}$ is a manifold is open and dense, and therefore contains C^∞ (and analytic) metrics.*

This result makes little difference in the proof of Donaldson's Theorem, but it is aesthetically satisfying.

Our proof of the Transversality Theorem (3.17) is valid for all $SU(2)$ bundles. We do not use the Chern class c_2, the intersection form ω, or the dimension of M in the proof, although we do rely on the low dimensionality of $SU(2)$. In fact, this proof also applies to $SO(3)$ bundles. An easy corollary for $U(1)$ bundles is

COROLLARY 3.21. *If ω is indefinite, then for an open dense set of metrics, there are no line bundle solutions to the self-dual or anti-self-dual equations.*

PROOF: Theorem 3.4 goes over word for word, as does the analogue of (3.16). The only difference is the index calculation in (3.17). For line bundles, the elliptic complex (3.18) reduces to

$$(3.22) \qquad\qquad 0 \to \Omega^0 \xrightarrow{d} \Omega^1 \xrightarrow{d_-} \Omega^2_- \to 0,$$

since the adjoint bundle is trivial. The index of (3.22) is computed in [**AHS**] to be $\frac{1}{2}(\chi - \tau)$, where χ is the Euler characteristic and τ is the signature. Denote by b_- the dimension of the maximal space on which ω is negative definite. The index works out to be $(1 + b_-)$. (That this is the "analytical" index follows directly from Poincaré duality.) But $h^0 = 1$ (the constant functions), and we are interested in $h^1 - h^2 = -b_-$ (cf. the proof of (3.17)). When $b_- > 0$ the Sard–Smale Theorem implies that generically there are no self-dual solutions. The nonexistence of anti-self-dual solutions follows from $b_+ > 0$ by reversing the underlying orientation and repeating the proof.

Of course, there are still solutions to the Yang–Mills equations in this case by Hodge–de Rham Theory. So the indefinite case provides an example where the topological lower bound in §2 is not attained.

§4 CONES ON $\mathbb{CP}^2$

We proved in §3 that for a dense set of metrics the irreducible connections $\hat{\mathcal{M}}$ in the moduli space form a smooth manifold. Now we examine the singular points $\{p_1, p_2, \ldots, p_m\} \subseteq \mathcal{M}$ corresponding to reducible connections. We show that after a small perturbation of $\mathcal{M}$, made either by hand or through a perturbation of the metric, a neighborhood of each singular point is homeomorphic to an open cone on $\mathbb{CP}^2$. Furthermore, these homeomorphisms are smooth off the singular points.

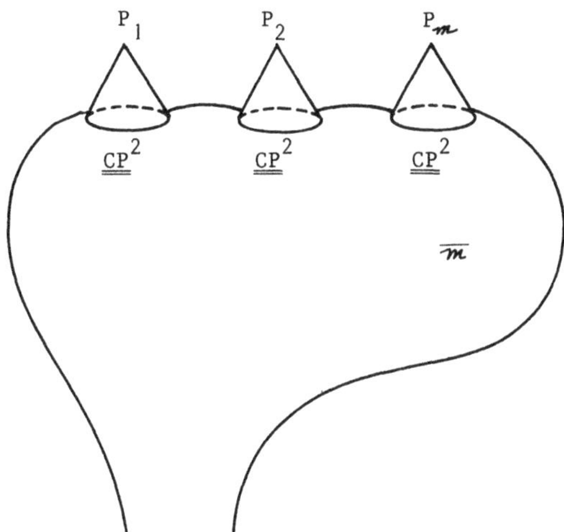

In particular, this proves the existence of irreducible self-dual connections near reducible solutions. By removing the open neighborhoods and attaching m copies of $\mathbb{CP}^2$ to the resulting gaps, we obtain half the cobordism (2.23) between M and $\amalg^m \mathbb{CP}^2$. It remains to prove that $\overline{\mathcal{M}}$ is orientable and $\partial\overline{\mathcal{M}} = M$; this will be done in §5–§9. Because arguments in this section closely parallel those of §3, we will often omit details (such as Sobolev subscripts).

Slices Again.

To begin we recall the basic properties of reducible connections. From (3.1) we know that $\mathcal{G}/\mathbf{Z}_2$ acts freely on irreducible connections $\hat{\mathcal{A}}$. At a reducible connection D the stabilizer $\mathcal{G}_D \subseteq \mathcal{G}$ is isomorphic to S^1. We will often find it more convenient to consider the quotient $S_D = \mathcal{G}_D/\mathbf{Z}_2 \simeq S^1$, since S_D acts freely on neighboring irreducible connections. The Lie algebra of $\mathcal{G}_D$ is the one dimensional kernel of

$$(4.1) \qquad D : \Omega^0(\operatorname{ad}\eta) \to \Omega^1(\operatorname{ad}\eta).$$

This represents the zeroth cohomology H_D^0 of the by now familiar elliptic complex

$$(4.2) \qquad 0 \to \Omega^0(\operatorname{ad}\eta) \xrightarrow{D} \Omega^1(\operatorname{ad}\eta) \xrightarrow{P_- D} \Omega_-^2(\operatorname{ad}\eta) \to 0.$$

Since (4.2) is $\mathcal{G}_D$ equivariant, $\mathcal{G}_D$ acts on all of the cohomology groups. The action on H_D^0 is trivial. In (4.9) we determine the action on H_D^1 and H_D^2. (Henceforth we identify elements in cohomology with their harmonic representatives.)

The slices

$$\mathcal{X}_D = \{A \in \Omega^1(\operatorname{ad}\eta) : D^*A = 0\}$$

give local charts for $\mathcal{X}$ near irreducible connections (3.2). For the reducible case we must take into account the S^1 symmetry. Recall the isotropy subgroup $\mathcal{G}_D = \{t = \exp\theta u : Du = 0, |u| = 1, 0 \leq \theta \leq 2\pi\}$. It acts on $\mathcal{X}_D$ by

$$(4.3) \qquad A \mapsto t^{-1}At$$

since $Dt = 0$. Moreover, the natural action of $\mathcal{G}$ on its Lie algebra is by conjugation

$$s \mapsto t^{-1}st.$$

In any Sobolev metric on $\Omega^i(\operatorname{ad}\eta)_{\ell-1}$ constructed using the covariant differentiation, $\mathcal{G}_D$ acts orthogonally, since t is pointwise unitary and parallel.

THEOREM 4.4. *Let D be a reducible connection and $\overline{D}$ its orbit in $\mathcal{X}$. Then for a sufficiently small neighborhood $\mathcal{O}_{\overline{D}}$ of $\overline{D}$ in $\mathcal{X}$, there is a local homeomorphism*

$$\mathcal{O}_{\overline{D}} \simeq \mathcal{X}_D/\mathcal{G}_D$$

which is smooth off the singular point $\overline{D} \in \mathcal{O}_{\overline{D}}$.

PROOF: We copy the proof of (3.2). Define

$$L_D : \Omega^1(\operatorname{ad}\eta) \times \mathcal{G} \to \Omega^0(\operatorname{ad}\eta)$$

$$\langle A, s \rangle \mapsto D^*(s^{-1}Ds + s^{-1}As).$$

Now the partial differential $\delta_2 L_D = D^*D$ in the direction s has a kernel H_D^0. Restrict to the L^2 orthogonal complement $H_D^{0\perp}$, i.e., the image of D^* in $\Omega^0(\operatorname{ad}\eta)$; then the second partial of the restriction

$$\tilde{L}_D : \Omega^1(\operatorname{ad}\eta) \times \exp(H_D^{0\perp}) \to H_D^{0\perp}$$

is invertible at $\langle 0, id \rangle$. Therefore, the equation $\tilde{L}_D(\tilde{A}, s) = 0$ has a unique solution $s = \exp u \in \exp(H_D^{0\perp})$ for $\tilde{A} \in \Omega^1(\operatorname{ad}\eta)$ small, and the solution $A = f(\tilde{A})$ to the equations

$$A = s^{-1}Ds + s^{-1}\tilde{A}s$$

$$(4.5)$$

$$D^*A = 0$$

depends smoothly on $\tilde{A}$. Now the adjoint action of $t \in \mathcal{G}_D$ on H_D^0 is trivial (since $Dt = 0$), so $\operatorname{ad} t$ maps $H_D^{0\perp}$ into itself. The same conclusion holds on the group level: conjugation by t maps $\exp(H_D^{0\perp})$ into itself. Thus if A satisfies (4.5), then also

$$t^{-1}At = (t^{-1}st)^{-1}D(t^{-1}st) + (t^{-1}st)^{-1}(t^{-1}\tilde{A}t)(t^{-1}st),$$

$$D^*(t^{-1}At) = 0.$$

It follows that $f(\operatorname{ad} t \cdot \tilde{A}) = \operatorname{ad} t \cdot f(\tilde{A})$, and from this equivariance we obtain $\mathcal{X}_D/\mathcal{G}_D \cong \mathcal{O}_D/\mathcal{G}$. (We omit the argument that $\mathcal{X}_D/\mathcal{G}_D$ injects into the quotient; it is similar to that in §3.)

Structure of the Singular Point.

The moduli space locally near a singular point is the kernel of

$$P_-F : \mathcal{X}_D \to \Omega^2_-(\operatorname{ad}\eta)$$

$$(4.6)$$

$$A \mapsto P_-(DA + A \wedge A)$$

divided by the action of $\mathcal{G}_D$. The linearization of P_-F at $A \in \mathcal{X}_D$ is the map P_-D in the complex (4.2). As a consequence of ellipticity, P_-D restricted to $\mathcal{X}_D$ is a (linear) Fredholm map. Hence P_-F is a (nonlinear) Fredholm map (cf. the discussion following (3.17)). The following "finiteness property" is the major step in proving the Sard–Smale Theorem [K], [Sm], [D].

LEMMA 4.7. *Suppose $\Psi : X \to Y$ is a (nonlinear) Fredholm map between Hilbert spaces, $\Psi(0) = 0$, and let $T = (d\Psi)_0 : X \to Y$ be the differential at the origin. Then there are splittings*

$$X = \operatorname{Ker} T \oplus X',$$

$$Y = \operatorname{Im} T \oplus Y',$$

and a map

$$\phi : X \to Y'$$

so that Ψ is equivalent near the origin to $T + \phi$ by a diffeomorphism of X. It follows that $\phi(0) = 0$ and $(d\phi)_0 = 0$. Furthermore, if a group G acts by an orthogonal action on X and Y, and Ψ is G equivariant, then X' and Y' are G invariant and ϕ is G equivariant.

In other words, a Fredholm map is locally equivalent to the sum of a linear map and a nonlinear map with finite dimensional range. The Sard–Smale Theorem is obtained by applying the ordinary Sard Theorem to the finite dimensional part ϕ. In our case, the situation is greatly simplified because the singularity is already known to be isolated.

PROOF: Since T is Fredholm, $\operatorname{Ker} T$ is finite dimensional and $\operatorname{Im} T$ is closed of finite codimension, and the orthogonal complements X' and Y' are well-defined. By the open mapping theorem $T|_{X'}$ has a unique bounded inverse $T^{-1} : \operatorname{Im} T \to X'$. Extend T^{-1} to Y by $\tilde{T} = T^{-1}(1 - \pi_C)$ where $\pi_C : Y \to Y'$ is the (finite dimensional) orthogonal projection onto the cokernel T. Then, if $\pi_K : X \to \operatorname{Ker} T$ is the orthogonal projection onto the kernel of T,

$$\chi : X \to X$$

$$x \mapsto \pi_K(x) + \tilde{T}(\Psi(x))$$

is a local diffeomorphism at 0. We have

$$(T \circ \chi)(x) = (1 - \pi_C)(\Psi(x)),$$

and defining

$$\phi(y) = \pi_C(\Psi(\chi^{-1}(y)))$$

gives the desired equivalence $\Psi \circ \chi^{-1} = T + \phi$. That $\phi(0)$ and $(d\phi)_0$ vanish is clear. The equivariance of ϕ follows from the equivariance of Ψ, T, and the naturality of the construction.

Apply this to the map (4.6) which specifies the anti-self-dual curvature on the slice through D. The linearization at zero is $P_- D|_{\mathrm{Ker}\, D^*}$, which is Fredholm since (4.2) is elliptic. Furthermore, $P_- F$ is $\mathcal{G}_D$ equivariant, and since $\mathcal{G}_D$ acts by rotation it follows that the moduli space near $\overline{D}$ is $(P_- D|_{\mathrm{Ker}\, D^*} + \overline{\phi})^{-1}(0)/\mathcal{G}_D$ for some $\mathcal{G}_D$ equivariant finite dimensional mapping

$$\overline{\phi} : \mathrm{Ker}\, D^* \to (\mathrm{Im}\, P_- D)^{\perp}.$$

We have proved

COROLLARY 4.8. *Let D be a reducible self-dual connection. Then near $\overline{D}$ the moduli space is homeomorphic to $\phi^{-1}(0)/\mathcal{G}_D$ for some $\mathcal{G}_D$ equivariant local map*

$$\phi : H_D^1 \to H_D^2.$$

Of course, H_D^1 and H_D^2 are the first and second cohomology of (4.2), identified with the harmonic spaces in that complex. Also, $\phi^{-1}(0)/\mathcal{G}_D$ is locally diffeomorphic to the moduli space except at the origin.

Since $(d\phi)_0 = 0$, we need $H_D^2 = 0$ to guarantee that $\phi^{-1}(0)$ is (locally) a manifold. Note that this is the ordinary implicit function theorem. The next proposition gives the structure of H_D^1, H_D^2 explicitly. We state it in more generality than necessary to illustrate our topological hypotheses.

PROPOSITION 4.9. *For a reducible connection D there are real isomorphisms*

$$H_D^1 \simeq \mathbf{C}^q$$

$$H_D^2 \simeq \mathbf{C}^p \oplus P_- H_{DR}^2(M)$$

under which $S_D \simeq S^1$ acts by the standard action on $\mathbf{C}^q$, $\mathbf{C}^p$ and by the trivial action on $P_- H_{DR}^2(M)$. If $P_- H_{DR}^2(M) = 0$, then $q = p + 3$.

PROOF: We first determine fixed points of the S_D action on H_D^1 and H_D^2. Let $u \in \Omega^0(\mathrm{ad}\,\eta)$, $|u| = 1$, generate the one dimensional kernel of D. If $A \in H_D^1$ is a fixed point, then $[u, A] = 0$, from which $A = \theta \otimes u$ for some $\theta \in \Omega^1$. By an argument similar to (3.15), we can deduce $d^* \theta = P_- d\theta = 0$. Thus $d\theta$ is self-dual, and by Stokes' Theorem,

$$0 = \int_M d(\theta \wedge d\theta) = \int_M d\theta \wedge d\theta = \int_M d\theta \wedge *d\theta = \int_M |d\theta|^2 * 1,$$

from which $d\theta = 0$. It follows that θ is harmonic, and since we assume $H^1_{DR}(M) = 0$, we conclude that $A = 0$. Similarly, a fixed point in H^2_D takes the form $\alpha \otimes u$, where $\alpha \in \Omega^2_-$ is harmonic, whence the fixed point set is identified with $P_- H^2_{DR}(M)$, the anti-self-dual second cohomology of M. Note that if the intersection form of M is positive definite, then $P_- H^2_{DR}(M) = 0$.

Any linear representation space V of S^1 can be decomposed as $V = V_{\mathbf{C}} \oplus V_{\mathbf{R}}$ where S^1 acts trivially on the real space $V_{\mathbf{R}}$ and by a direct sum of one-dimensional representations on the complex space $V_{\mathbf{C}}$. Furthermore, the only linear irreducible action of S^1 which is free away from the origin is the standard representation on $\mathbf{C}$. These observations imply decompositions

$$H^1_D \simeq \mathbf{C}^q \oplus V_1$$

$$H^2_D \simeq \mathbf{C}^p \oplus V_2$$

for some real spaces V_1, V_2, and our previous discussion identifies $V_1 = 0$ and $V_2 = P_- H^2_{DR}(M)$. Finally, the Atiyah–Singer Index Theorem applied to (4.2) fixes $q = p + 3$ when $P_- H^2_{DR}(M) = 0$.

Now we can identify the local structure of $\mathcal{M}$ near the singular points when the second cohomology vanishes.

COROLLARY 4.10. *If $H^2_D = 0$, then near the origin $\phi^{-1}(0)/\mathcal{G}_D$ is homeomorphic to a cone on $\mathbf{CP}^2$, diffeomorphic except at the vertex.*

PROOF: First, $H^1_D \simeq \mathbf{C}^3$ and $S_D \simeq S^1 \subset \mathbf{C}$ acts by complex multiplication. Corollary 4.8 and the Implicit Function Theorem identify $\phi^{-1}(0)/S_D = \mathbf{C}^3/S^1$ near the origin, where S^1 acts by multiplication by units in $\mathbf{C}$. Since $\mathbf{CP}^2 = (\mathbf{C}^3 \backslash \{0\})/\mathbf{C}^*$, the orbit space $\mathbf{C}^3/S^1 = \mathbf{CP}^2 \times [0,1)/\langle x,0 \rangle \sim \langle x',0 \rangle$ is an open cone on $\mathbf{CP}^2$. Finally, $\phi^{-1}(0)/\mathcal{G}_D = \phi^{-1}(0)/S_D$ since the center $\mathbf{Z}_2 \subset \mathcal{G}_D$ acts trivially.

With these preliminaries it is easy to construct our perturbed moduli space by hand [**D**]. Notice that $P_- H^2_{DR}(M) = 0$, as we assume now that M has positive definite intersection form.

THEOREM 4.11. *There exists a perturbation of $\mathcal{M}$ so that locally about a reducible self-dual connection, $\mathcal{M}$ is homeomorphic to an open cone on $\mathbf{CP}^2$. The identification is a diffeomorphism off the vertex.*

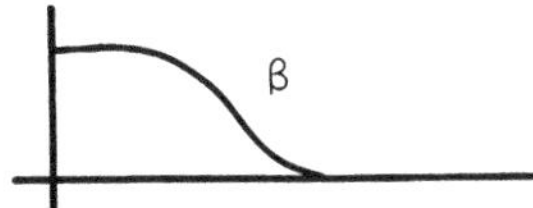

PROOF: It suffices to consider each singular point separately. We perturb $\phi : \mathbf{C}^{p+3} \to \mathbf{C}^p$. Let L be a surjective linear map $L : \mathbf{C}^{p+3} \to \mathbf{C}^p$ and β a cutoff function.

Define $\tilde{\phi} : \mathbf{C}^{p+3} \to \mathbf{C}^p$ by

$$\tilde{\phi}(z) = \phi(z) + \varepsilon^2 \beta \left(\frac{|z|}{\varepsilon} \right) L(z)$$

for a suitably small $\varepsilon > 0$. Then $\tilde{\phi}$ is $\mathcal{G}_D$ equivariant, and $(d\tilde{\phi})_0 = L$ is surjective. So in a small neighborhood of the origin, $\tilde{\phi}^{-1}(0)$ is a smooth 6-dimensional manifold. The perturbed moduli space is $\tilde{\phi}^{-1}(0)/S^1$ near the origin, and the result follows by the arguments in (4.10).

In the next section we give an alternative approach to the genericity question.

Perturbing the Metric.

Fix a base metric g on M. We study the cohomology $H^2_{D_i}$ at each of the singular connections $D_1, \ldots, D_m$ (which arise from splittings $\eta = \lambda \oplus \lambda^{-1}$), and prove that the cohomology groups vanish for a generic metric. Then (4.10) gives the local structure of $\mathcal{M}$ near $\overline{D}_i$. Set $h_i^2 = \dim H^2_{D_i}$, and fix i. For $D = D_i$ we have the rank one curvature $F = F_D = \sigma \otimes u$, $\sigma \in \Omega^2_+$, $u \in \Omega^0(\mathrm{ad}\,\eta)$, $|u| = 1$. By the Bianchi identity,

$$0 = DF = d\sigma \otimes u + \sigma \wedge Du,$$

from which

$$(4.12) \qquad\qquad d\sigma = Du = 0$$

as in (3.15); in other words, σ is harmonic and u is parallel. Recall that $D : \Omega^0(\mathrm{ad}\,\eta) \to \Omega^1(\mathrm{ad}\,\eta)$ has a one dimensional kernel, which in this case is spanned by u, and so we can split

$$\mathrm{ad}\,\eta = \mathbf{R} \cdot u \oplus \zeta$$

orthogonally. Then D acts trivially on the line bundle $\mathbf{R} \cdot u$, and together with the induced connection, $\zeta \oplus_{\mathbf{R}} \mathbf{C} \simeq \eta$. Therefore, we can split the complex (4.2) into the direct sum of the ordinary anti-self-dual de Rham complex

$$(4.13) \qquad 0 \to \Omega^0 \xrightarrow{d} \Omega^1 \xrightarrow{d_-} \Omega^2_- \to 0$$

and the complex

$$(4.14) \qquad 0 \to \Omega^0(\zeta) \xrightarrow{\mathcal{D}} \Omega^1(\zeta) \xrightarrow{\mathcal{D}} \Omega^2_-(\zeta) \to 0.$$

Here $d_- = P_- d$ and $\mathcal{D} = P_- D$ are notations we adopt henceforth. The positive definiteness of the intersection form implies that (4.13) has index 1, and this cohomology is represented by the constant functions in Ω^0. (If $P_- H^2_{DR}(M) \neq 0$, then the proof of Corollary 3.21 implies that generically there are no singular points in the moduli space, so that there is no contribution to H^2_D from (4.13) in this case, too.) Thus we study the complex (4.14), which we note has index -6, and show that $h^2 = \dim \operatorname{Coker} \mathcal{D}$ vanishes for a dense set of metrics. Our previous transversality theorem does not apply, since we are exactly in the case where F has rank one and $(F, \Phi) = 0$ in (3.4). However, our present genericity problem is easier, since the m discrete solutions vary smoothly in the metric, and we need not worry about complicated solution sets.

Our first lemma implies that the set of metrics for which $h^2 = 0$ is open.

LEMMA 4.15. *The operators D and $\mathcal{D}$ depend smoothly on the metric g, and $h^2(g)$ is upper semicontinuous:*

$$\overline{\lim_{g \to g_0}} \, h^2(g) \leq h^2(g_0).$$

Of course, $\mathcal{D}$ is only determined up to a gauge transformation, so that smoothness is really measured in the orbit space $\mathcal{X}$.

PROOF: The smooth dependence follows from the fact that zero is a regular value of $P_- F$ on the line bundle λ (cf. (3.4)). Furthermore, the space $\Omega^2_-(\zeta)$ varies smoothly with the metric. Now $h^2(g)$ is the dimension of the kernel of $\mathcal{D}\mathcal{D}^* : \Omega^2_-(\zeta) \to \Omega^2_-(\zeta)$, and since this operator is elliptic, it has discrete spectrum, whence $h^2(g)$ is upper semicontinuous (just as for finite dimensional matrices; the proof for elliptic (thus Fredholm) operators follows immediately from (4.7)).

Next, we prove that in some sense the first cohomology of (4.14)—more precisely, its image under D—lies generically in directions (in M) not "parallel" to the curvature. Note that if $A \in \Omega^1(\zeta)$ is exact, $A = Dv$, then

$$DA = D^2 v = [F, v] = [\sigma \otimes u, v] = \sigma \otimes [u, v]$$

is parallel to F. The converse holds for generic metrics.

LEMMA 4.16. *There exists an open dense set of C^k metrics such that if $A \in \Omega^1(\zeta)$ satisfies $\mathcal{D}A = 0$, and if $DA = \sigma \otimes w$ on the open set where $F \neq 0$, then $A = Dv$ for some $v \in \Omega^0(\zeta)$.*

PROOF: The elliptic equation

$$D^*(A - Dv) = 0$$

has a unique solution $v \in \Omega^0(\zeta)$, since D has no kernel in ζ. Then for $\tilde{A} = A - Dv$,

$$
\begin{aligned}
D\tilde{A} &= DA - [F, v] \\
&= \sigma \otimes (w - [u, v]) \\
&= \sigma \otimes \tilde{w}
\end{aligned}
$$
(4.17)

is parallel to F, and $\tilde{A}$ is harmonic:

$$\mathcal{D}\tilde{A} = D^*\tilde{A} = 0.$$
(4.18)

We prove that for an open dense set of metrics, (4.17) and (4.18) imply that $\tilde{A} = 0$.

Openness follows easily since the existence of $\tilde{A} \neq 0$ satisfying (4.17) and (4.18) is a closed condition. To see this, note that σ, $\mathcal{D}$, and D^* all vary smoothly with the metric, and we can use the weak compactness of solutions for elliptic equations to extract a weakly convergent subsequence from a sequence of solutions.

To show that the set of "good" metrics is dense, we consider analytic metrics. Then σ, $\mathcal{D}$, and $\mathcal{D}^*$ are all analytic, and so the zero set of σ is well-behaved, in particular has measure zero. Since $[u, [u, \tilde{w}]] = -4\tilde{w}$ for elements of $\mathfrak{su}(2)$,

$$D\tilde{A} = [F, \tilde{v}],$$

where $4\tilde{v} = -[u, \tilde{w}]$. Thus

$$[F \wedge \tilde{A}] = D^2\tilde{A} = D[F, \tilde{v}] = [F \wedge D\tilde{v}].$$

But as in the proof of (3.4), the self-dual form σ acts invertibly as a map $\Omega^1 \to \Omega^3$, and since $u \notin \zeta$, F acts invertibly $\Omega^1(\zeta) \to \Omega^3(\zeta)$. Therefore, $\tilde{A} = D\tilde{v}$ on the set where $F \neq 0$. Moreover, $\tilde{v}$ is unique and analytic on this set, and $|\nabla^j \tilde{v}|$ is bounded by $|\nabla^{j-1}\tilde{A}|$. So $\tilde{v}$ can be extended across the zero set of F by power series expansions. Now $D^*\tilde{A} = D^*D\tilde{v} = 0$ globally, from which $\tilde{v} = 0$. Hence $\tilde{A} = 0$, and $A = Dv$.

Now we proceed to our main result.

THEOREM 4.19. *For an open dense set of C^k metric, h^2 vanishes at each singular point in the moduli space.*

PROOF: Since finite intersections of open dense sets are open and dense, it suffices to consider each reducible connection separately. Furthermore, we can restrict to the open dense set in (4.16). Although our present linear setup is considerably simpler than the nonlinear situation in §3, we still make use of our transversality machinery as we have already introduced it.

Consider the map

$$\mathcal{Q} : \Omega^1(\zeta)\backslash\{0\} \times \mathcal{C} \to \Omega^0(\zeta) \oplus \Omega^2_-(\zeta)$$

$$\langle A, \varphi \rangle \mapsto \langle D^*((\varphi^{-1})^*A), P_-((\varphi^{-1})^*DA) \rangle.$$

Here $\mathcal{C}$ is the space of C^k frame changes (cf. (3.4)), and the operator D is defined implicitly as the self-dual solution relative to the metric φ^*g. That said, $\mathcal{Q}^{-1}(0)$ is the nonzero first cohomology $H^1_D(\zeta)$ relative to φ^*g. If we show that zero is a regular value of $\mathcal{Q}$, then it follows that for a generic metric $H^1_D(\zeta)$ is a 6 dimensional linear space. (The projection $\mathcal{Q}^{-1}(0) \to \mathcal{C}$ has index 6, and we can apply Sard–Smale as in (3.17).) For these metrics $h^2 = 0$.

Suppose $\langle v, \Phi \rangle \in \text{Coker}\, \delta\mathcal{Q}_{\langle A,\varphi \rangle}$, where $\mathcal{Q}(A,\varphi) = 0$. Infinitesimal variations $r \in \mathfrak{c}$ in the Lie algebra of $\mathcal{C}$, which preserve the curvature ($r^*F = 0$) fix D to first order, and so for these variations

$$\delta_2\mathcal{Q}_{\langle A,\varphi \rangle}(r) = \langle D^*((\varphi^{-1})^*(r^*A)), P_-((\varphi^{-1})^*(r^*DA)) \rangle.$$

Then setting $\tilde{\Phi} = \varphi^*(\Phi)$, and denoting the adjoint of D in the metric $\varphi^* g$ by D^*, we obtain

$$(4.20) \qquad\qquad Dv = 0$$

$$(4.21) \qquad\qquad D^* \tilde{\Phi} = 0,$$

$$(4.22) \qquad\qquad (r^* DA, \tilde{\Phi})_{\varphi^* g} = 0,$$

as in (3.13), (3.14). Since D has no kernel in ζ, it follows from (4.20) that $v = 0$. Also, Lemma 4.16 implies that where $DA = \sigma \otimes w$ aligns with the curvature and $F \neq 0$, A vanishes. But we exclude $A \equiv 0 \in \Omega^1(\zeta)$ from our domain, and by unique continuation (6.38), A and F do not vanish on open sets. Hence there is an open set U on which $DA \neq \sigma \otimes w$ and $F \neq 0$. Fix a frame in which $\sigma = \sigma^{13}$; this is possible by (3.5). Then

$$F = \sigma^{13} \otimes u,$$

$$DA = \sum_{i=2}^{4} \sigma^{1i} \otimes w_i,$$

$$\tilde{\Phi} = \sum_{j=2}^{4} \alpha^{1j} \otimes \Phi_j,$$

and either w_2 or w_4 is nonzero, say $w_2 \neq 0$. Using the frame changes (3.8) and (3.10) in (4.22), and noting that $r^* F = 0$, we conclude that $(w_2, \Phi_j) = 0$. So $\tilde{\Phi}$ has rank one in U since ζ is two dimensional. By (4.21) and the argument in (3.15), we have $\tilde{\Phi} = 0$ on U. By unique continuation $\tilde{\Phi}$ vanishes everywhere, so that zero is a regular value of $\mathcal{Q}$, and the theorem follows.

Donaldson's Theorem relies on the invariance of signature under *oriented* cobordism. We prove here that $\hat{\mathcal{M}}$, the smooth manifold used to construct the cobordism $M \sim \mathrm{II}^m \mathbf{CP}^2$, is orientable.

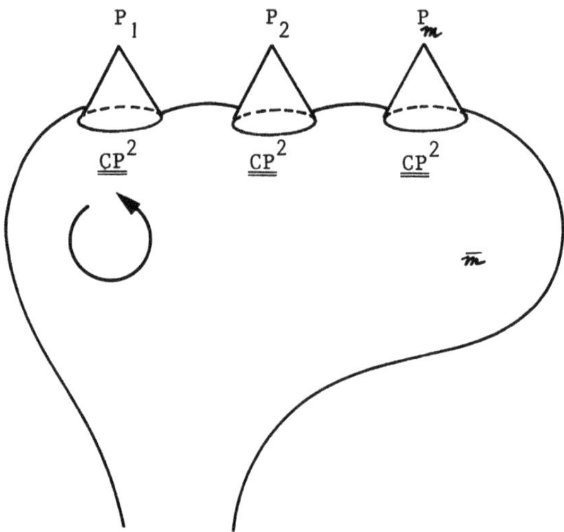

Now $\hat{\mathcal{M}} \subseteq \hat{\mathcal{X}}$ and we show that $T\hat{\mathcal{M}}$ is orientable by producing an orientable extension ξ of $T\hat{\mathcal{M}}$. The bundle ξ is

$$
\begin{array}{ccc}
T\hat{M} & \hookrightarrow & \xi \\
\downarrow & & \downarrow \\
\hat{\mathcal{M}} & \subseteq & \hat{\mathcal{X}}
\end{array}
$$

the equivariant *index bundle* of the non-linear elliptic complex $\{D^*, P_-F\}$. That ξ is orientable follows from the simple connectivity of $\hat{\mathcal{X}}$, and the proof that $\pi_1(\hat{\mathcal{X}}) = 0$ constitutes the bulk of our work. We first show that up to a $\mathbf{Z}_2$, $\pi_1(\hat{\mathcal{X}})$ is the set of homotopy classes $[M, S^3]$. This can be done by homotopy theory, or more geometrically by using the Pontrjagin–Thom identification of $[M, S^3]$ with classes of framed cobordant framed submanifolds of M. This second computation is carried out in Appendix B. Finally, we account for the $\mathbf{Z}_2$ and thereby complete the proof that ξ is orientable.

Index Bundles.

The index bundle of a parametrized family of Fredholm operators between real Hilbert spaces generalizes the numerical index of a linear Fredholm map. If $T : \mathcal{H}_1 \to \mathcal{H}_2$ is a linear Fredholm map between Hilbert spaces, then

$$(5.1) \qquad \operatorname{ind} T = \dim \operatorname{Ker} T - \dim \operatorname{Coker} T \in \mathbf{Z}$$

defines the index. In the parametrized case the index $\operatorname{ind} T$ is a *virtual bundle* over the parametrizing manifold X, i.e., an element of $KO(X)$. Recall that under direct sum the equivalence classes of real vector bundles over X form a semigroup whose Grothendieck group completion is $KO(X)$. An element $\xi \in KO(X)$ can be written as a formal difference $\xi = \xi' - \xi''$ of two real bundles over X. Of course, the virtual bundle ξ is stably equivalent to an honest vector bundle if X is compact. A real bundle ξ over X is orientable iff the first Stiefel–Whitney class $w_1(\xi) \in H^1(X; \mathbf{Z}_2)$ vanishes, and for $\xi = \xi' - \xi'' \in KO(X)$ we define $w_1(\xi) = w_1(\xi') - w_1(\xi'')$ and decree ξ orientable iff $w_1(\xi) = 0$. Note that this definition depends on the additivity of w_1. If ξ is actually a bundle ($\xi'' = 0$), then the two notions of orientability agree. Also, if $\pi_1(X) = 0$, then any (virtual) bundle over X is orientable. On the category of *compact* manifolds, the functor KO admits a representation convenient for elliptic operator theory. Namely, let $\mathcal{H}$ be a separable, infinite dimensional real Hilbert space. Then there is an isomorphism

$$(5.2) \qquad \operatorname{ind} : [X, \operatorname{Fred}(\mathcal{H})] \xrightarrow{\sim} KO(X),$$

where $\operatorname{Fred}(\mathcal{H})$ is the space of Fredholm operators on $\mathcal{H}$ and $[X, \operatorname{Fred}(\mathcal{H})]$ denotes the group of homotopy classes of maps $x \to \operatorname{Fred}(\mathcal{H})$ [**A1**, Appendix]. Let $\psi : X \to \operatorname{Fred}(\mathcal{H})$. The basic idea is to define a virtual bundle $\operatorname{ind}\psi$ over X by analogy with (5.1). The definition

$$(5.2) \qquad (\operatorname{ind}\psi)_x = \operatorname{Ker}\psi(x) - \operatorname{Coker}\psi(x)$$

is correct if $\operatorname{Ker}\psi$ and $\operatorname{Coker}\psi$ have constant rank for all $x \in X$. In general, though, $\dim \operatorname{Ker}\psi(x)$ and $\dim \operatorname{Coker}\psi(x)$ are not locally constant, so that $\operatorname{Ker}\psi(x)$ and $\operatorname{Coker}\psi(x)$ do not glue together to form vector bundles. However, by an argument similar to (4.7), we can show that for a small perturbation of ψ (which is allowed since we are only interested in ψ up to homotopy), equation (5.3) does make sense. Intuitively, then, we think of the index bundle as the stable class of the difference of the kernel and cokernel. For $X = pt$ we recover (5.1). Unfortunately, no direct modification of (5.3) gives a well-defined global index bundle if X is noncompact—the infinite sum of local finite dimensional perturbations is not finite dimensional. One can resolve this problem by *defining* $KO(X) = [X, \mathrm{Fred}(\mathcal{H})]$ for X noncompact [**P**, §18]. This is not so useful to us, although the idea is basic to our construction. Alternatively, consider only restrictions to compact submanifolds of X. This suffices for our purposes since orientability can be determined by looking only at compact sets. The generalization of the index bundle to the case where the Hilbert space $\mathcal{H}_x$ varies with $x \in X$ is treated in [**AS**].

To begin we remove the singular points of our moduli space $\mathcal{M}$, which cannot affect the orientability question. We have been looking at a canonical slice, by which we associate to the tangent space of $\hat{\mathcal{M}}$ at $\overline{D}$ the five dimensional linear subspace of $\Omega^1(\mathrm{ad}\,\eta)$

$$E_D = \{A \in \Omega^1(\mathrm{ad}\,\eta) : D^*A = 0, P_- DA = 0\}$$

for any lift D of $\overline{D}$. The vector bundle

$$E_{\mathcal{SD}} = \{A \in E_D, D \in \mathcal{S}\hat{\mathcal{D}}\}$$

forms a five dimensional bundle over $\mathcal{S}\hat{\mathcal{D}}$, with elements of $\mathcal{G}$ acting as linear bundle transformations, covering the action of $\mathcal{G}$ on the base irreducible self-dual connections $\mathcal{S}\hat{\mathcal{D}}$. Canonically,

$$(5.4) \qquad\qquad T\hat{\mathcal{M}} = E_{\mathcal{SD}}/\mathcal{G}$$

since the tangent space at the orbit $\overline{D}$ is E_D. Now we extend this entire operation as best we can to $\hat{\mathcal{X}}$. Define

$$L(D) = D^* \oplus P_- D : \Omega^1(\mathrm{ad}\,\eta) \to \Omega^0(\mathrm{ad}\,\eta) \oplus \Omega^2_-(\mathrm{ad}\,\eta).$$

This is an elliptic, and therefore Fredholm, operator for each $D \in \hat{\mathcal{A}}$. Let $\mathcal{H}_1 = \Omega^1(\mathrm{ad}\,\eta)$ and $\mathcal{H}_2 = \Omega^0(\mathrm{ad}\,\eta) \oplus \Omega^2_-(\mathrm{ad}\,\eta)$. (Remember these are Hilbert spaces; we are suppressing the Sobolev subscripts.) Now

$$(5.5) \qquad\qquad L : \hat{\mathcal{A}} \times \mathcal{H}_1 \to \hat{\mathcal{A}} \times \mathcal{H}_2$$

is a parametrized family of elliptic operators. Moreover, we allow $\mathcal{G}$ to act on $\mathcal{H}_1$ and $\mathcal{H}_2$ in the natural fashion. This turns out to be

$$\langle s, \phi \rangle \mapsto s^{-1} \phi s$$

for $s \in \mathcal{G}$, $\phi \in \Omega^i(\mathrm{ad}\,\eta)$. Then L is equivariant with respect to this action. The algebraic properties are easily checked, and we briefly describe the analytical correctness in Appendix A with Sobolev subscripts. Therefore, dividing out by $\mathcal{G}$ gives

$$(5.6) \qquad\qquad \overline{L} : E_1 \to E_2,$$

where $E_1 = (\hat{A} \times \mathcal{H}_1)/\mathcal{G}$ and $E_2 = (\hat{A} \times \mathcal{H}_2)/\mathcal{G}$ are both Hilbert space bundles over $\hat{\mathcal{X}} = \hat{A}/\mathcal{G}$. The operator $\overline{L}$ acquires from L the property of acting as a linear Fredholm operator of index 5 on each fiber over $\overline{D} \in \hat{\mathcal{X}}$. We define the virtual bundle

$$\xi = \mathrm{Ker}\,\overline{L} - \mathrm{Coker}\,\overline{L}.$$

By our previous discussion, $\xi \mid S \in KO(S)$ is well-defined on compact subspaces $S \subset \hat{\mathcal{X}}$—by a small perturbation $\xi \mid S$ is represented by the difference of two real vector bundles.

THEOREM 5.7. *If $\hat{\mathcal{X}} = \hat{A}/\mathcal{G}$ is simply connected, then $T\hat{\mathcal{M}}$ is orientable.*

PROOF: We show that the Stiefel–Whitney class $w_1(T\hat{\mathcal{M}})$ vanishes. If not, there exists a circle $\Gamma \subseteq \hat{\mathcal{M}}$ such that $w_1(T\hat{\mathcal{M}}|_\Gamma) \neq 0$. Let S be a compact set in $\hat{\mathcal{X}}$ containing Γ so that $i_*[\Gamma] = 0$, where $i : \Gamma \hookrightarrow S$ is the inclusion, and $[\Gamma]$ is the homology class of Γ. For example, we can take S to be the homotopy of Γ to a point in the simply connected manifold $\hat{\mathcal{X}}$. Now

$$w_1(T\hat{\mathcal{M}}|_\Gamma) = w_1(\xi|_\Gamma) = i^*(w_1(\xi|_S)).$$

Evaluating on the homology class $[\Gamma]$,

$$w_1(T\hat{\mathcal{M}}|_\Gamma)[\Gamma] = i^*(w_1(\xi|_S))[\Gamma]$$
$$= w_1(\xi|_S)(i_*[\Gamma])$$
$$= 0.$$

This contradiction proves that $w_1(T\hat{\mathcal{M}}) = 0$.

Components of $\mathcal{G}$.

At this point we remind the reader that the gauge group $\mathcal{G}$ has an ineffective $\mathbf{Z}_2$ in its action on $\hat{\mathcal{A}}$, which can be thought of as the center $\{1, -1\}$ of $SU(2)$. These elements of the center describe elements of $\mathcal{G}$ because they are invariant under the adjoint action of $SU(2)$, which is used to construct $\mathcal{G} = C^\infty(\operatorname{Aut}\eta)$.

Let $\tilde{\mathcal{G}} = \mathcal{G}/\mathbf{Z}_2$. Now $\tilde{\mathcal{G}} \to \hat{\mathcal{A}} \to \hat{\mathcal{X}}$ is a principal fibration. Here the action of $\tilde{\mathcal{G}}$ on $\hat{\mathcal{A}}$ at D_0 is

$$\langle s, A \rangle \mapsto s^{-1}D_0 s + s^{-1}As$$

as usual. (Some authors reverse the action. In our case, $s \in \mathcal{G}$ acts on $\Omega^i(\operatorname{ad}\eta)$ by $\operatorname{ad} s^{-1}$; it is probably more common for the action to be $\operatorname{ad} s$. We repeat this makes little difference, but, like the sign of the Laplace operator, it can be confusing.) As in finite dimensions, this fibration gives rise to an exact sequence in homotopy

$$\cdots \to \pi_1(\hat{\mathcal{A}}) \to \pi_1(\hat{\mathcal{X}}) \to \pi_0(\tilde{\mathcal{G}}) \to \pi_0(\hat{\mathcal{A}}) \to \pi_0(\hat{\mathcal{X}}).$$

Recall that exact sequences of *pointed* sets (like $\pi_0(\hat{\mathcal{A}})$) are well-defined. Now $\mathcal{A}$ is contractible, and since the space $\mathcal{A}\backslash\hat{\mathcal{A}}$ of reducible connections has infinite codimension in $\mathcal{A}$ (there is an infinite parameter family of local perturbations that render a reducible connection irreducible), $\hat{\mathcal{A}}$ has the weak homotopy type of a point [**S**, Theorem 2]. Hence

$$(5.8) \qquad\qquad \pi_1(\hat{\mathcal{X}}) = \pi_0(\tilde{\mathcal{G}}).$$

We first compute $\pi_0(\mathcal{G})$. A simple consequence of the classification of $SU(2)$ bundles by second Chern class, which will be evident in the grafting procedure of §6, is that all such bundles can be constructed from two patches. In other words, there is an open cover $M = M^+ \cup M^-$ with $M^+ \simeq B^4$ (the 4-ball), $M^+ \cap M^- \simeq S^3 \times (0,1)$, and a clutching function $h : M^+ \cap M^- \to SU(2)$ so that the principal $SU(2)$ bundle P associated to η is

$$P = M^+ \times SU(2) \amalg M^- \times SU(2)/\sim$$

where

$$(5.9) \qquad \langle m^+, g \rangle \sim \langle m^-, g' \rangle \quad \text{iff} \quad m^+ = m^- \text{ and } g' = h(m^+)g.$$

Since $c_2(\eta)[M] = -1$, for any $t \in (0,1)$ the map $x \mapsto h(x,t)$ has degree -1, and modifying by a homotopy we can assume that $h(x,t) = x^{-1}$. Let $B \subseteq M^+$ be a smaller 4-ball, and to simplify matters assume that $B = M^+ \backslash M^-$.

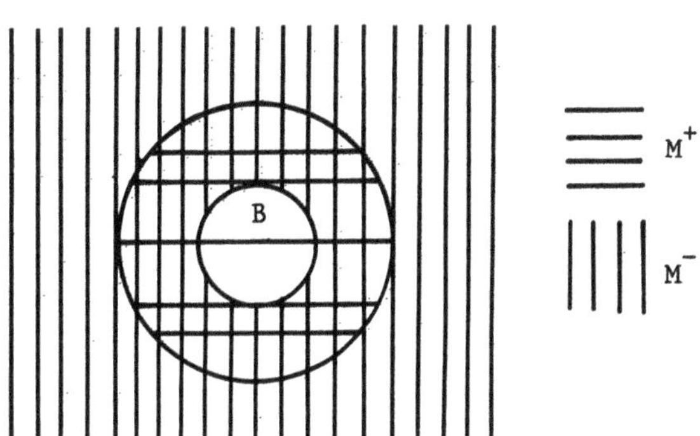

Define

$$\mathcal{G}_0 = \{s \in \mathcal{G} : s|_B \equiv 1\}.$$

LEMMA 5.10. *The inclusion* $\iota : \mathcal{G}_0 \hookrightarrow \mathcal{G}$ *induces an isomorphism*

$$\iota_* : \pi_0(\mathcal{G}_0) \to \pi_0(\mathcal{G}).$$

PROOF: Restricted to B, a gauge transformation is just a map $B \to SU(2)$. Any such map is homotopic to the constant map $B \to \{1\}$ since $SU(2)$ is connected. Hence ι_* is onto. Now suppose that $s \in \mathcal{G}_0$ is in the component of the identity in $\mathcal{G}$. Then there is a path in $\mathcal{G}$ from s to 1, and restricting to B this is a loop of maps $B \to SU(2)$ in which the base point is sent to the constant map $B \to \{1\}$. But $SU(2)$ is simply connected, and so we have a homotopy to the constant loop over B, that is, a path in $\mathcal{G}_0$ from s to 1, whence ι_* is injective.

Given $s \in \mathcal{G}_0$, s can be described from (5.9) as the pair of maps

$$\begin{aligned}
s^- &: M^- \to SU(2) \\
s^+ &: M^+ \to SU(2)
\end{aligned}$$

(5.11)

with

$$s^-(x,t) = x^{-1} s^+(x,t) x \quad \text{on} \quad S^3 \times (0,1) = M^- \cap M^+.$$

However, $s \equiv 1$ on $M \setminus M^- = B$, so $s^-(x,0) = s^+(x,0) = 1$. Regard

$$s^- \in [(M^-, \partial M^-), (SU(2), 1)]$$
$$= [(M, pt), (S^3, pt)]$$
$$= [M, S^3],$$

where at the last stage we use the fact that S^3 is simply connected.

This proves

PROPOSITION 5.12. $\pi_0(\mathcal{G}) = \pi_0(\mathcal{G}_0) = [M, S^3]$.

We briefly outline what we need to know about $[M, S^3]$. A more geometric argument can be found in Appendix B. Given any degree one map $\sigma : M \to S^4$, there is a pullback $\sigma^* : [S^4, S^3] \to [M, S^3]$.

PROPOSITION 5.13. *If the intersection form ω is odd, then $[M^4, S^3] = 0$. If it is even $[M^4, S^3] = \mathbb{Z}_2$. Moreover, if ω is even, $\mathbb{Z}_2$ is the image of $[S^4, S^3] = \mathbb{Z}_2$ under σ^*.*

PROOF: By the Steenrod Classification Theorem [**Sp**],

$$[M, S^3] = H^4(M, \mathbb{Z}_2) / \operatorname{Image} \tilde{S}q^2.$$

If ω is the bilinear form in the middle dimension, $\tilde{S}q^2 : H^2(M, \mathbb{Z}) \to H^4(M, \mathbb{Z}_2)$ is just composition of the usual Steenrod square with reduction $\mod 2 : \tilde{S}q^2(\alpha) = \omega(\alpha, \alpha) \mod 2$. This gives the result. Moreover, the construction is functorial, which implies $[M^4, S^3] \subseteq \sigma^*([S^4, S^3])$ as required.

COROLLARY 5.14. $\pi_0(\mathcal{G}) = \begin{cases} \mathbb{Z}_2 & \text{if } \omega \text{ is even,} \\ 0 & \text{if } \omega \text{ is odd.} \end{cases}$

We briefly describe the generator of $[S^4, S^3]$, which is the suspension of the Hopf map. Let $S^1 = \{e^{i\Lambda\theta}, \Lambda \in \mathfrak{su}(2)\}$. The Hopf map $H : S^3 = SU(2) \to S^2 = SU(2)/S^1$ is just the map to left cosets:

$$H(x) = [x] = \{xe^{i\Lambda\theta}\}.$$

Its suspension $\Sigma H : \Sigma S^3 = S^4 \to \Sigma S^2 = S^3$ is simply

$$(5.15) \qquad\qquad \Sigma H(x, \varphi) = \langle [x], \varphi \rangle,$$

where $0 \leq \varphi \leq \pi$ is the polar angle in S^4 and S^3.

The Element -1.

From the fibration

$$\mathbf{Z}_2 \xrightarrow{j} \mathcal{G} \to \tilde{\mathcal{G}}$$

we obtain the exact sequence of pointed sets

$$(5.16) \qquad \cdots \to \pi_0(\mathbf{Z}_2) \xrightarrow{j_*} \pi_0(\mathcal{G}) \to \pi_0(\tilde{\mathcal{G}}) \to 0.$$

If ω is odd, (5.8) and (5.14) combine to prove that $\pi_1(\hat{\mathcal{X}}) = 0$. We restrict our attention to even ω. Then (5.16) becomes

$$(5.17) \qquad \mathbf{Z}_2 \xrightarrow{j_*} \mathbf{Z}_2 \to \pi_0(\tilde{\mathcal{G}}) \to 0,$$

and $\pi_0(\tilde{\mathcal{G}}) = 0$ iff j_* maps $-1 \in \mathbf{Z}_2 \subseteq \mathcal{G}$ to the nontrivial element of $\pi_0(\mathcal{G})$. Our analysis is based on the explicit construction leading to (5.12).

PROPOSITION 5.18. *j_* is onto.*

PROOF: Fix the circle subgroup $S^1 = \{e^{i\Lambda\theta}\} \subseteq SU(2)$ as before, and define $\lambda(t) = e^{i\pi t\Lambda}$, $0 \le t \le 1$, to be a path from $+1$ to -1 in S^1. We push $-1 \in \mathcal{G}$ to $\mathcal{G}_0$ using λ to obtain $s \in \mathcal{G}_0$ defined by (cf. (5.11))

$$s^+ = \begin{cases} \lambda(t) & \text{on } M^+ \cap M^- \\ +1 & \text{on } D \end{cases},$$

$$s^- = \begin{cases} x^{-1}\lambda(t)x & \text{on } M^+ \cap M^- \\ -1 & \text{on } M^- \backslash M^+ \end{cases}$$

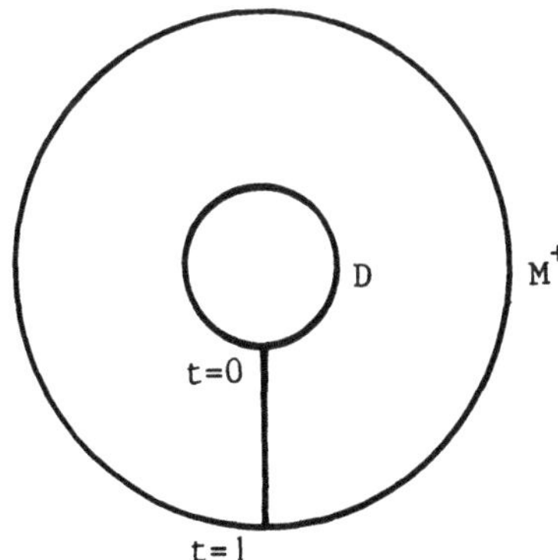

Then $s \in \mathcal{G}_0$ by (5.11) and is in the same path component of $\mathcal{G}$ as -1.

In the identification of $\pi_0(\mathcal{G})$ with $[M^4, S^3]$ in (5.12), s is represented by s^- extended to be $+1$ on B. Moreover, if we define $\sigma : M^4 \to S^4$ by

$$\sigma(M^+ \backslash M^-) \;=\; \text{north pole } (\varphi = 0),$$

$$\sigma(M^- \backslash M^+) \;=\; \text{south pole } (\varphi = \pi),$$

$$\sigma|_{M^+ \cap M^-} : S^3 \times [0,1] \to \Sigma S^3 \text{ by projection,}$$

we see that $s^- = u \circ \sigma$, where $u : S^4 \to S^3$ is given by

$$u(x, \varphi) = x^{-1} e^{i\varphi \Lambda} x.$$

Then u factors through the suspension $\overline{\Sigma H}$ of the inverse of the Hopf map:

$$
\begin{array}{ccccccc}
 & S^4 & \xrightarrow{\;u\;} & S^3 & \qquad \langle x, \varphi \rangle & \to & x^{-1} e^{i\varphi \Lambda} x \\
\overline{\Sigma H} \searrow & & \nearrow \theta & & \searrow & & \nearrow \\
 & \Sigma S^2 = S^3 & & & & \langle [x^{-1}], \varphi \rangle &
\end{array}
$$

Here θ is the diffeomorphism which identifies the great half circle through the point $[x^{-1}] \in SU(2)/U(1) = S^2$.

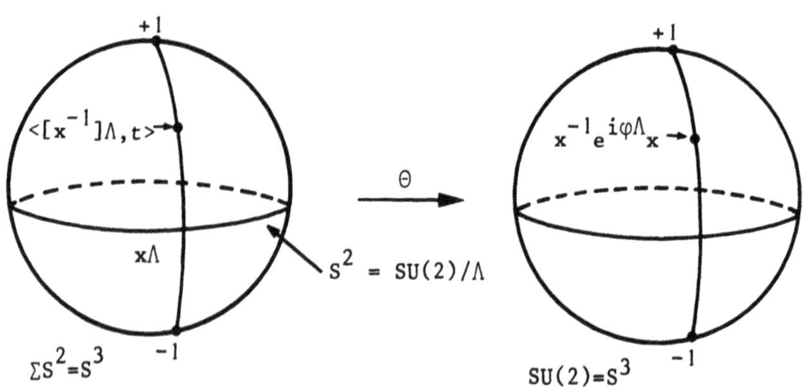

on the equator with the great half circle $\{x^{-1} e^{i\varphi \Lambda} x : 0 \le \varphi \le \pi\}$. By (5.13) and the discussion following (5.14), s^- generates the nontrivial element of $\mathbf{Z}_2 = \pi_0(\mathcal{G})$. So j_* in (5.14) is surjective, and $\pi_0(\tilde{\mathcal{G}}) = 0$. Finally, $\pi_1(\hat{\mathcal{X}}) = 0$ follows from (5.8). All in all, we have proved

THEOREM 5.19. $\hat{\mathcal{M}}$ is orientable.

At this stage our moduli space $\hat{\mathcal{M}}$, although by now a smooth orientable manifold, may still be empty, if there are no reducible connections! A theorem of Clifford Taubes [**T**] rules out this gloomy possibility. He establishes the existence of self-dual connections on a 4-manifold M whose intersection form is positive definite. Taubes' Theorem complements work of Atiyah, Hitchin, and Singer [**AHS**], who construct moduli spaces for a more restricted class of manifolds. For these "half-conformally flat" manifolds, twistor theory can be used to convert Yang–Mills into a problem in algebraic geometry. In particular, the self-dual Yang Mills equations are well understood on S^4 (with the standard metric), although the topology of the moduli space for $k > 2$ is not completely known. Our 4-manifold M is not in general half-conformally flat, and other methods are required. Taubes uses analytic techniques to build self-dual connections on M from the solutions on S^4. The $k = 1$ instantons on S^4 have a *center* $b \in S^4$ and a *scale* $\lambda \in \mathbf{R}^+$. As $\lambda \to 0$ the instanton becomes localized near b. One can imagine a limiting connection at $\lambda = 0$ whose curvature is supported at b. Taubes grafts the localized self-dual connections onto M, where they pick up a small anti-self-dual curvature, and for λ sufficiently small he perturbs them

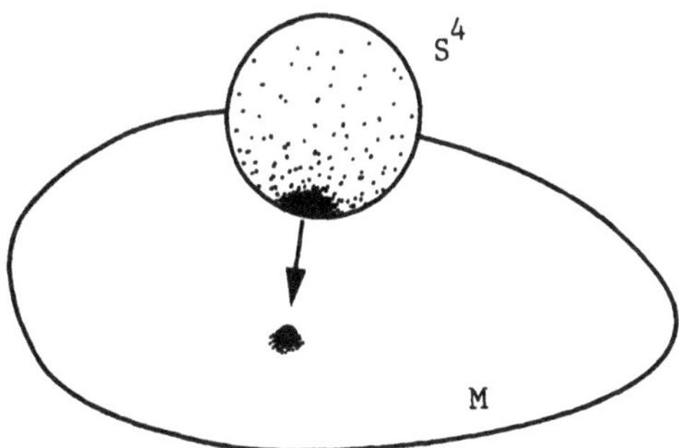

slightly to obtain self-dual connections. There results a family of instantons on M, parametrized by $(0, \lambda_0) \times M$, and in a later chapter we prove that these essentially form a collar of M in $\overline{\mathcal{M}}$, the limiting connections $\lambda = 0$ being adjoined to form the compactification $\overline{\mathcal{M}} = \mathcal{M} \cup M$.

We begin this chapter by presenting the solutions for the $k = 1$ bundle on S^4. Then we describe the grafting procedure and derive curvature estimates which demonstrate that the grafted connections are almost self-dual. The rest of §6 is devoted to statements of analytic results—some standard, some specialized—that we need in §7–§9. In the next chapter we complete the proof of Taubes' Theorem by annihilating the anti-self-dual part of the curvature using a small perturbation.

Both our presentation of the solutions on S^4 and our derivation of the Weitzenböck formulas in Appendix C stress the role of a group of symmetries. Principal bundles are appropriate here as the symmetry group is built into their geometry, and we do not hesitate to use them. However, we promised in §2 not to *rely* on principal bundles, and since the geometry of principal bundles may be somewhat unfamiliar to analysts, we wish to point out that only formulas (6.7) and (6.8) from the first section are used in a significant way. Alternative derivations of these formulas may be found in the physics literature [**BPST**], [**JNR**]. Nevertheless, our description of the moduli space for the sphere will provide good intuition for the general case. Furthermore, the Weitzenböck formulas can be derived by a more straightforward computation in normal coordinates. At the other extreme, we hope that the inclusion of some standard results from PDE will benefit those whose previous experience in this area is limited.

Instantons on S^4.

Further information about solutions on S^4 can be found in Atiyah's monograph [**A2**]; we follow his exposition closely. In particular, we use the quaternionic notation

$$x = x^1 + x^2 i + x^3 j + x^4 k \in \mathbf{H}$$

for elements of $\mathbf{R}^4$. Of course, $i^2 = j^2 = k^2 = ijk = -1$ and i, j, k anticommute. Recall that

$$\operatorname{Re} x = x^1$$

$$\operatorname{Im} x = x^2 i + x^3 j + x^4 k$$

$$\overline{x} = x^1 - x^2 i - x^3 j - x^4 k,$$

and

$$|x|^2 = x\overline{x}$$

is the usual norm on $\mathbf{R}^4$. The Lie algebra $\operatorname{Im}\mathbf{H}$ of imaginary quaternions is isomorphic to $\mathfrak{su}(2)$, and the Lie group $Sp(1)$ of unit quaternions is isomorphic to $SU(2)$. We compute differentials as usual, although extreme caution is advised due to the noncommutativity of the quaternions. Self-duality is especially easy in this notation: the coefficients of i, j, k in

(6.1)
$$dx \wedge d\overline{x} = -2\{(dx^1 \wedge dx^2 + dx^3 \wedge dx^4)i + (dx^1 \wedge dx^3 - dx^2 \wedge dx^4)j$$
$$+(dx^1 \wedge dx^4 + dx^2 \wedge dx^3)k\}$$

form a basis for self-dual 2-forms, and the coefficients of

$$d\overline{x} \wedge dx = 2\{(dx^1 \wedge dx^2 - dx^3 \wedge dx^4)i + (dx^1 \wedge dx^3 + dx^2 \wedge dx^4)j$$
$$+(dx^1 \wedge dx^4 - dx^2 \wedge dx^3)k\}$$

form a basis for anti-self-dual 2-forms. Both forms take values in $\mathfrak{su}(2)$, and since under the identification $\mathfrak{su}(2) \simeq \operatorname{Im}\mathbf{H}$, the inner product (2.12) on $\mathfrak{su}(2)$ becomes $(x, y) = 2\operatorname{Re}(x\overline{y})$ on $\operatorname{Im}\mathbf{H}$, their norms are

$$\|dx \wedge d\overline{x}\|^2 \,\mathrm{vol} = 2\operatorname{Re}\{(dx \wedge d\overline{x}) \wedge {*\overline{(dx \wedge d\overline{x})}}\}$$

(6.2)
$$= 48\,\mathrm{vol}$$

$$= \|d\overline{x} \wedge dx\|^2 \,\mathrm{vol}.$$

Endow $\mathbf{H}^2$ with the standard quaternionic inner product

$$((\langle p^1, p^2\rangle, \langle q^1, q^2\rangle)) = p^1\overline{q}^1 + p^2\overline{q}^2.$$

Then $\operatorname{Re}(\ ,\ )$ is the standard real inner product on $\mathbf{R}^8 \simeq \mathbf{H}^2$. The vectors of real norm one form the 7-sphere S^7, and this fibers over $S^4 \simeq \mathbf{HP}^1$ by

$$\pi : \langle q^1, q^2\rangle \mapsto [q^1, q^2], \qquad |q^1|^2 + |q^2|^2 = 1.$$

Here $\mathbf{HP}^1 = \{[q^1, q^2]\}$ is the projectivization of $\mathbf{H}^2$—$[q^1, q^2] = [pq^1, pq^2]$ in $\mathbf{HP}^1$ for all nonzero $p \in \mathbf{H}$. The fiber of π is $SU(2) = \{p \in \mathbf{H} : |p| = 1\}$ which acts by conjugate left multiplication on S^7, so that $\pi : S^7 \to S^4$ is a principal $SU(2)$ bundle (whose associated complex 2-plane bundle is denoted η_{S^4} in the sequel), in fact the $k = 1$ bundle. (One can verify this last statement by computing the Chern number using (2.9) and (6.6).)

Since $Sp(2)$ acts by isometries on S^7, the real orthogonal complement to the vertical tangent space

$$VTS^7_{\langle q^1,q^2\rangle} = \{\langle pq^1, pq^2\rangle : p \in \operatorname{Im}\mathbf{H}\}$$

defines a homogeneous horizontal distribution on S^7. It is easy to verify that this is a connection and

$$(6.3) \qquad \Theta = \operatorname{Im}(q^1 d\bar{q}^1 + q^2 d\bar{q}^2)$$

is the corresponding connection form. The formula for the curvature Ω is complicated, but at the point $\langle q_1, q_2\rangle = \langle 0, 1\rangle$ it takes the simple form

$$(6.4) \qquad \Omega_{\langle 0,1\rangle} = dq^1 \wedge d\bar{q}^1.$$

The horizontal subspace at $\langle 0, 1\rangle$ is spanned over $\mathbf{H}$ by $\frac{\partial}{\partial q^1}$, and so in view of (6.1), $\Omega_{\langle 0,1\rangle}$ is the pullback from $\mathbf{HP}^1$ of a self-dual form. Now the $Sp(2)$ homogeneity implies that the curvature is self-dual everywhere. Furthermore, this $Sp(2)$ action on S^7, which preserves the connection, projects onto the usual $SO(5)$ action on S^4. The conformal invariance of the self-dual Yang–Mills equation (6.40) shows that the double cover $SL(2, \mathbf{H})$ of the larger group $SO(5, 1) \simeq SL(2, \mathbf{H})/\{\pm 1\}$ of conformal transformations acts on η_{S^4} to give other instantons. Much more difficult is the fact that *all* $k = 1$ instantons are thus obtained [**AHS**]. Granting this, we see that the moduli space for S^4 is the hyperbolic ball $B^5 = SL(2, \mathbf{H})/Sp(2) = SO(5, 1)/SO(5)$.

We obtain local versions of (6.3) and (6.4) from the stereographic projection

$$\mathbf{R}^4 \simeq \mathbf{H} \to \mathbf{HP}^1 \simeq S^4$$

$$x \mapsto [x, 1]$$

and the section

$$\mu(x) = \frac{\langle x, 1\rangle}{\sqrt{1 + |x|^2}}$$

of $\eta_{S^4}|_{\mathbf{R}^4}$. Then

$$(6.5) \qquad A = \mu^*\Theta = \operatorname{Im}\left(\frac{x d\bar{x}}{1 + |x|^2}\right),$$

$$(6.6) \qquad F = \mu^*\Omega = \frac{dx \wedge d\bar{x}}{(1 + |x|^2)^2}.$$

More explicit formulas are obtained by expanding (6.5) and (6.6). Write

$$A = \sum_{i=1}^{4} A_i dx^i,$$

$$F = \sum_{i<j} F_{ij} dx^i \wedge dx^j;$$

then

$$A_1 = \mathrm{Im}\left(\frac{x}{1+|x|^2}\right) = \frac{x^2 i + x^3 j + x^4 k}{1+|x|^2},$$

$$A_2 = \mathrm{Im}\left(\frac{-xi}{1+|x|^2}\right) = \frac{-x^1 i + x^3 k - x^4 j}{1+|x|^2},$$

$$A_3 = \mathrm{Im}\left(\frac{-xj}{1+|x|^2}\right) = \frac{-x^1 j - x^2 k + x^4 i}{1+|x|^2},$$

$$A_4 = \mathrm{Im}\left(\frac{-xk}{1+|x|^2}\right) = \frac{-x^1 k + x^2 j - x^3 i}{1+|x|^2}.$$

The factor 48 in

$$k = 1 = \frac{1}{8\pi^2} \int_{\mathbf{R}^4} \frac{48 dx}{(1+|x|^2)^2}$$

is explained by the factor of 2 in the trace inner product, e.g.

$$i \in \mathbf{H} \leftrightarrow \begin{pmatrix} i & 0 \\ 0 & -i \end{pmatrix} \in \mathfrak{su}(2) \quad \text{and} \quad -tr\begin{pmatrix} i & 0 \\ 0 & -i \end{pmatrix}^2 = 2,$$

and the factor of 2 in each of the 6 curvature terms, e.g.

$$F_{12} = \frac{-2i}{(1+|x|^2)^2},$$

which gets squared when we take norms.

Conformal transformations of the (local) form

$$T_{\lambda,b} : x \mapsto \lambda(x - b), \qquad \lambda > 0, b \in S^4$$

parametrize the moduli space $SO(5,1)/SO(5)$ if we identify $T_{\lambda,b} \sim T_{1/\lambda,b^*}$, where b^* is the point antipodal to b. The instanton $T_{\lambda,b}^* \Theta$ is said to have *center b* and *scale λ*. (Our basic instanton Θ is centerless of unit scale.) We

explain the terminology as follows. The homothety $T_\lambda = T_{\lambda,0}$ is represented
on $\mathbf{R}^4$ by $x \mapsto \lambda x$, and

$$(6.7) \qquad \mu^* T_\lambda^* \Theta = \mathrm{Im}\left(\frac{x\,d\bar{x}}{\lambda^2 + |x|^2}\right),$$

$$(6.8) \qquad \mu^* T_\lambda^* \Omega = \frac{\lambda^2 dx \wedge d\bar{x}}{(\lambda^2 + |x|^2)^2}.$$

Notice that as $\lambda \to 0$ in (6.8), the curvature concentrates at $x = 0$. More
generally the curvature of $T_{\lambda,b}^* \Theta$ is centered at b, the spread being determined by the scale λ. From this we obtain a vivid description of the moduli
space B^5. The center of the ball

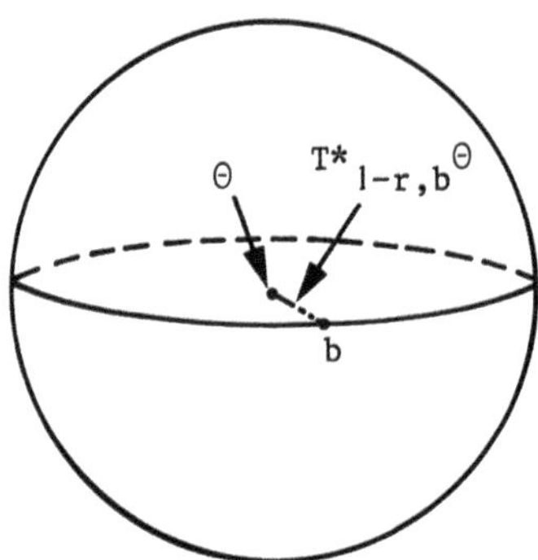

is the basic instanton Θ, and the point along the radius to $b \in S^4 = \partial B^5$
at distance r from the center is the instanton $T_{1-r,b}^* \Theta$. As we approach the
boundary point b, the curvature becomes increasingly concentrated at b.
It is only natural, then, to compactify B^5 by attaching the boundary S^4.
Intuitively, points of S^4 represent self-dual connections whose curvature is
a "δ-form", i.e., a form supported at a single point. The boundary of the
compactified moduli space is the original manifold S^4, and there is a collar

$$[0, \lambda_0) \times S^4 = \{T_{\lambda,b}^* \Theta : \lambda < \lambda_0\} \cup S^4, \qquad \lambda_0 < 1.$$

A Grafting Procedure.

The preceding description of the moduli space near its boundary generalizes to any 4-dimensional manifold M satisfying the hypotheses of (2.25).

We construct connections on M by grafting on the S^4 instantons. That these are approximately self-dual will be made precise in (6.11). In §7 we will make a perturbation to annihilate the anti-self-dual part of the curvature, and in §9 we deduce the product structure of the end of $\mathcal{M}$. Now any $SU(2)$ bundle η on M can be constructed as the pullback of an $SU(2)$ bundle on S^4 via a map $\Phi : M \to S^4$. If Φ has degree 1, then $\eta = \Phi^*(\eta_{S^4})$ is the $k = 1$ bundle on M. Let $\rho(M)$ be the injectivity radius of M, i.e., the radius of the smallest normal coordinate system on M. For each $y \in M$ and $\lambda \in \left(0, \frac{\rho(M)^2}{4}\right)$ we construct a degree 1 map

$$\Phi_{\lambda,y} : M \to S^4$$

as follows. Let U be a neighborhood of y which is mapped diffeomorphically onto an open set in $\mathbf{R}^4$ by $\exp^{-1}$. Fix a C^∞ cutoff function $\beta : [0, \infty] \to [0, 1]$ satisfying

$$
\begin{aligned}
\beta(r) &= 1, && 0 \le r \le 1; \\
\beta(r) &= 0, && r \ge 2;
\end{aligned}
$$

(6.9)

β monotone decreasing;

$$|\beta'(r)| \le 2.$$

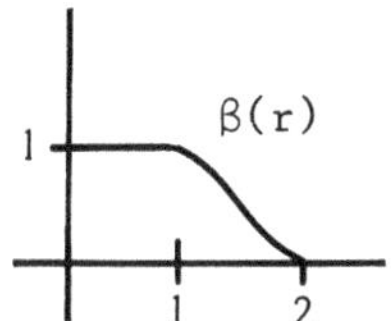

Then define $\varphi_{\lambda,y} : M \to S^4$ by

$$
(6.10) \qquad \varphi_{\lambda,y}(x) =
\begin{cases}
\dfrac{\exp^{-1} x}{\beta\left(\dfrac{|\exp^{-1} x|}{\sqrt{\lambda}}\right)}, & x \in U, \\[4ex]
\infty, & x \notin U.
\end{cases}
$$

(The factor $\sqrt{\lambda}$ is chosen with 20–20 hindsight!) Here we identify

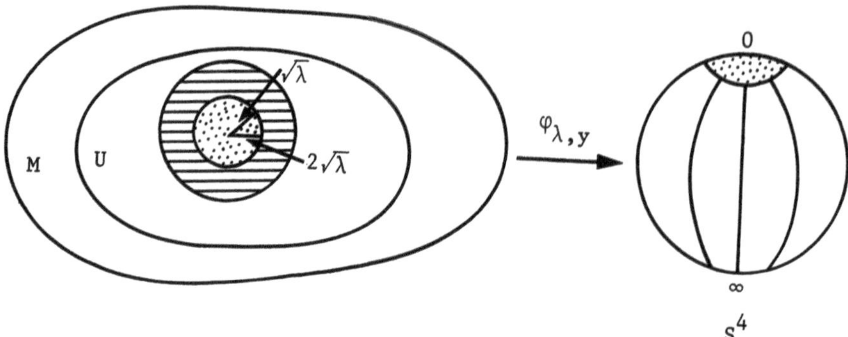

$\mathbf{R}^4 \cup \{\infty\} = S^4$ by stereographic projection. The ball $\exp(B_{\sqrt{\lambda}}) \subseteq M$ is mapped onto $B_{\sqrt{\lambda}} \subseteq \mathbf{R}^4$ by $\varphi_{\lambda,y}$, whereas the annulus $\exp(B_{2\sqrt{\lambda}} \backslash B_{\sqrt{\lambda}}) \subseteq M$ is stretched onto the exterior of $B_{\sqrt{\lambda}} \subseteq \mathbf{R}^4$, all the way out to ∞. Define $\Phi_{\lambda,y}$ as the composition

$$\Phi_{\lambda,y} : M \xrightarrow{\varphi_{\lambda,y}} S^4 \xrightarrow{T_\lambda} S^4,$$

where T_λ is the dilation with scale factor λ. Then $\eta = \Phi^*_{\lambda,y}(\eta_{S^4})$ is the $k = 1$ bundle on M, and η inherits the connection $D_\lambda = \Phi^*_{\lambda,y}(\Theta)$ with curvature $F_\lambda = \Phi^*_{\lambda,y}(\Omega)$. On $M \backslash U$ the curvature vanishes, and on U, F_λ is given explicitly in normal coordinates by

$$(6.11) \qquad F_\lambda(x) = \frac{\lambda^2}{\left(\lambda^2 + \dfrac{|x|^2}{\beta\left(\frac{|x|}{\sqrt{\lambda}}\right)^2}\right)^2} \, d\left(\frac{x}{\beta\left(\frac{|x|}{\sqrt{\lambda}}\right)}\right) \wedge d\left(\frac{\overline{x}}{\beta\left(\frac{|x|}{\sqrt{\lambda}}\right)}\right)$$

(cf. (6.8), (6.10)). The following estimates quantify the statement that F_λ is almost self-dual.

THEOREM 6.12. *For each p satisfying $1 < p \leq \infty$ there exist constants $c_1(p), c_2(p)$ independent of λ such that*

    (i) $\|F_\lambda\|_{L^p} \leq c_1(p)\lambda^{4/p-2}$,

    (ii) $\|P_- F_\lambda\|_{L^p} \leq c_2(p)\lambda^{2/p}$.

As before, P_- is the projection onto anti-self-dual 2-forms. The norm $\|\ \|_{L^p}$ is defined by

$$\|F\|_{L^p} = \left(\int_M |F|^p\right)^{1/p}$$

From now on we delete the volume form in the integral when this will cause no confusion. Also, we number the constants $c_1, c_2, \ldots$ for clarity. Observe that (6.12) holds without any global hypotheses on M.

PROOF: Let g_x be the metric of M expressed in normal coordinates x, and let δ denote the standard metric on $\mathbf{H}$ given by $|x|$. Then elementary properties of the exponential map imply $g_0 = \delta$, $\frac{dg_x}{dx}\big|_{x=0} = 0$, so that a Taylor expansion yields

$$(6.13) \qquad |g_x - \delta| \leq c_3 |x|^2$$

where c_3 is a constant depending on the curvature of M near y. Restrict $\lambda < \frac{1}{8c_3}$ so that $\|g_x - \delta\| \leq \frac{1}{2}$ in $B_{2\sqrt{\lambda}}$. Estimates can be made using δ instead of g_x at the expense of a constant. Now pointwise projection onto the anti-self-dual part of a 2-form, viewed as a map $\{\text{metrics on } \mathbf{R}^4\} \to \mathrm{End}(\Lambda^2 \mathbf{R}^4)$, is continuous. From (6.13),

$$(6.14) \qquad |P_- F_\lambda(x) - F_\lambda^-(x)| \leq c_4 |x|^2,$$

where F_λ^- is the anti-self-dual part of F_λ with respect to δ.

We work first on $B_{\sqrt{\lambda}}$. Here

$$F_\lambda(x) = \frac{\lambda^2 dx \wedge d\bar{x}}{(\lambda^2 + |x|^2)^2}$$

is self-dual with respect to δ. Then (6.2) and (6.14) give

$$|P_- F_\lambda(x)| \leq c_5 \frac{\lambda^2 |x|^2}{(\lambda^2 + |x|^2)^2} \leq \frac{1}{4} c_5,$$

so that

$$(6.15) \qquad \left(\int_{B_{\sqrt{\lambda}}} |P_- F_\lambda(x)|^p \right)^{1/p} \leq c_5 \lambda^{2/p}.$$

Now

$$|F_\lambda(x)| \leq c_7 \frac{\lambda^2}{(\lambda^2 + |x|^2)^2},$$

and the bound

$$\int_0^{\sqrt{\lambda}} \frac{\lambda^{2p} r^3 dr}{(\lambda^2 + r^2)^{2p}} \leq \lambda^{4-2p} \int_0^\infty \frac{s^3 ds}{(1 + s^2)^{2p}}$$

shows that

$$(6.16) \qquad \left(\int_{B_{\sqrt{\lambda}}} |F_\lambda(x)|^p \right)^{1/p} \leq c_8(p) \lambda^{4/p - 2}.$$

Consider the annulus $B_{2\sqrt{\lambda}} \backslash B_{\sqrt{\lambda}}$. Expanding (6.11),

$$F_\lambda(x) = \frac{\beta\left(\frac{r}{\sqrt{\lambda}}\right) 4\lambda^2}{\left(\beta\left(\frac{r}{\sqrt{\lambda}}\right) 2\lambda^2 + r^2\right)^2} \left[\frac{dx \wedge d\overline{x}}{\beta\left(\frac{r}{\sqrt{\lambda}}\right)^2} \right.$$
$$\left. + \frac{\beta'\left(\frac{r}{\sqrt{\lambda}}\right)}{2\beta\left(\frac{r}{\sqrt{\lambda}}\right)^3 r\sqrt{\lambda}} \left[(dx)\overline{x} \wedge (dx)\overline{x} - x\, d\overline{x} \wedge x\, d\overline{x} \right] \right],$$

where $r = |x|$. Now $|(dx)\overline{x} \wedge (dx)\overline{x} - x\, d\overline{x} \wedge x\, d\overline{x}| = O(r^2)$, and there are only two terms to estimate. Then using (6.9) we can easily prove that for $\sqrt{\lambda} < r < 2\sqrt{\lambda}$ each of

$$\frac{\beta^2 \lambda^2}{(\beta^2 \lambda^2 + r^2)^2}, \qquad \frac{\beta\beta' \lambda^{3/2} r}{(\beta^2 \lambda^2 + r^2)^2}$$

is bounded independent of λ. (This is where the choice of $\sqrt{\lambda}$ as cutoff parameter is crucial.) Hence

$$(6.17) \qquad \left(\int_{B_{2\sqrt{\lambda}} \backslash B_{\sqrt{\lambda}}} |F_\lambda(x)|^p \right)^{1/p} \leq c_9 \lambda^{2/p},$$

and since $|P_- F_\lambda(x)| \leq |F_\lambda(x)|$, (6.17) also holds for $P_- F_\lambda$. Finally, collecting (6.15)–(6.17), and noting that $F_\lambda \equiv 0$ on $M \backslash B_{2\sqrt{\lambda}}$, we have (6.12).

Tools from Analysis.

We collect a few analytic facts that will be needed in §7–§9. More specifically, we state the Weitzenböck formulas, basic facts about Sobolev spaces, and some elementary properties of elliptic partial differential equations. Our discussion is brief, and we refer to standard texts (e.g. [**Au**], [**F**], [**GT**], [**Tr**]) for details.

To analyze a *twisted* function ψ—that is, a section of a vector bundle ξ— on any manifold M, it is often sufficient to study the norm $|\psi|$ and apply known results about *ordinary* functions on M. For these arguments the pointwise formula

$$(6.18) \qquad (\nabla^*\nabla\psi, \psi) = \frac{1}{2}\Delta|\psi|^2 + |\nabla\psi|^2 \geq \frac{1}{2}\Delta|\psi|^2$$

is a basic link between the trace Laplacian $\nabla^*\nabla$ on twisted functions and the Laplacian Δ on ordinary functions. (Of course, we postulate metrics on M, ξ and a metric compatible covariant derivative on ξ to make sense of (6.18).) We use the analyst's positive Laplacians which, with respect to an orthonormal frame $\{e_k\}$ of the tangent space for which the connection form vanishes at a point, are given at that point by

$$(6.19) \qquad \nabla^*\nabla f = \Delta f = -\sum_k \nabla_{e_k}\nabla_{e_k} f$$

on functions f. The opposite sign convention is also commonly used, and we hope that ours will not be the cause of undue confusion. We also use a first order analogue of (6.18) called *Kato's inequality*, which states that on the open set where $\psi \neq 0$,

$$(6.20) \qquad |\nabla\psi| \geq |d|\psi||$$

(the proof is any easy application of Cauchy–Schwartz). This gives a more delicate version of (6.18):

$$(6.21) \qquad (\nabla^*\nabla\psi, \psi) \geq |\psi|\Delta|\psi|.$$

The elliptic operators of primary importance for us form a complex

$$(6.22) \qquad 0 \to \Omega^0(\operatorname{ad}\eta) \xrightarrow{D} \Omega^1(\operatorname{ad}\eta) \xrightarrow{\sqrt{2}\mathcal{D}} \Omega^2_-(\operatorname{ad}\eta) \to 0,$$

and (6.18) indicates a need to relate the Laplacian of (6.22),

$$\begin{array}{ccc}
\Omega^0(\operatorname{ad}\eta) & \Omega^1(\operatorname{ad}\eta) & \Omega^2_-(\operatorname{ad}\eta) \\
\downarrow D^*D & \downarrow 2\mathcal{D}\mathcal{D}^* + DD^* & \downarrow 2\mathcal{D}\mathcal{D}^* , \\
\Omega^0(\operatorname{ad}\eta) & \Omega^1(\operatorname{ad}\eta) & \Omega^2_-(\operatorname{ad}\eta)
\end{array}$$

to $\nabla^*\nabla$, the Laplacian of the covariant derivative operator

$$(6.23) \qquad \Omega^i(\operatorname{ad}\eta) \xrightarrow{\nabla} T^*M \otimes \Omega^i(\operatorname{ad}\eta), \qquad i = 0, 1, 2.$$

Here $\mathcal{D} = P_- D$ as in §4. Such a relation exists because the symbols of (6.22) and (6.23) agree. (This explains the factor of 2 in (6.22).) A derivation of the following *Weitzenböck formulas* is deferred to Appendix C.

$$(6.24) \qquad D^* D = \nabla^* \nabla \qquad \text{on } \Omega^0(\operatorname{ad} \eta),$$

$$(6.25)$$
$$2\mathcal{D}^* \mathcal{D} + D D^* = \nabla^* \nabla + \operatorname{Ric}(\cdot) - 2[\cdot \lrcorner P_+ F] \qquad \text{on } \Omega^1(\operatorname{ad} \eta),$$

$$(6.26) \qquad 2\mathcal{D}\mathcal{D}^* = \nabla^* \nabla - 2W^-(\cdot) + \frac{R}{3} + [P_- F, \cdot] \qquad \text{on } \Omega^2_-(\operatorname{ad} \eta).$$

(Compare [**T**, §2], [**Pa**, §1], [**Bou**].) We explain the notation. In general, the Riemannian curvature decomposes into irreducible pieces $W \oplus \operatorname{Ric}_0 \oplus R$, where $W : \Omega^2 \to \Omega^2$ is the *Weyl curvature*, $\operatorname{Ric}_0 : \Omega^1 \to \Omega^1$ is the *traceless Ricci curvature*, and $R \in \mathbf{R}$ is the *scalar curvature*. The full Ricci tensor $\operatorname{Ric} = \operatorname{Ric}_0 \oplus R$. In four dimensions the Weyl curvature decomposes into self-dual and anti-self-dual pieces: $W = W^+ + W^-$ [**ST**]. The action of the curvature on forms extends trivially to forms with values in $\operatorname{ad} \eta$. As before, the curvature $F = P_+ F + P_- F$ of D decomposes. Also, there are $SO(4)$ equivariant bilinear maps

$$(6.27) \qquad\qquad\qquad \lrcorner : \Omega^1 \otimes \Omega^2_- \to \Omega^1$$

$$(6.28) \qquad\qquad\qquad [\,,\,] : \Omega^2_- \otimes \Omega^2_- \to \Omega^2_-$$

given by contraction and Lie bracket; for the latter we identify $\Omega^2_- \simeq \mathfrak{so}(3)$. Then the extrinsic curvature terms in (6.25) and (6.26) are defined by (6.27), (6.28), and the Lie bracket in $\operatorname{ad} \eta$. Lichnerowicz [**L**] and Bochner [**Bo**] used formulas like (6.24)–(6.26) to prove vanishing theorems. We will combine the Weitzenböck formulas with (6.18) and (6.21) to make estimates.

The formula

$$(6.29) \qquad \mathcal{D}^*(f\Phi) = f\mathcal{D}^*\Phi - df \lrcorner \Phi, \qquad f \in \Omega^0, \Phi \in \Omega^2_-(\operatorname{ad} \eta)$$

is useful; the minus sign is due to integration by parts.

Assume that M is *compact*. Recall (§3) that the Sobolev space $H_1(\xi)$ consists of L^2 sections ψ whose first derivative $\nabla\psi$ is also L^2. The expression

$$\|\psi\|^2_{H_1} = \int_M |\psi|^2 + |\nabla\psi|^2$$

94

defines the norm on $H_1(\xi)$. The *Sobolev Embedding Theorem* in four dimensions states that $H_1(\xi) \hookrightarrow L^4(\xi)$ continuously [**Tr**, §24]. In other words,

$$
(6.30) \qquad
\begin{aligned}
\left(\int_M |\psi|^4 \right)^{1/4} &= \|\psi\|_{L^4} \leq c_{10} \|\psi\|_{H_1} \\
&= c_{10} \left(\int |\psi|^2 + |\nabla\psi|^2 \right)^{1/2}
\end{aligned}
$$

for some constant c_{10} depending on M. To derive (6.30) from the Sobolev inequality for *ordinary* functions, use (6.20). Finally, for small $\varepsilon > 0$ an easy consequence of Rellich's Theorem is that $H_1 \hookrightarrow L^{4-\varepsilon}$ is a *compact* embedding. We stress that both this result and the inequality (6.30) depend on the compactness of the underlying manifold.

In §8 we will make use of more refined results about Sobolev spaces [**Au**], [**Ma**], [**P**]. When a vector bundle ξ over a compact Riemannian n-manifold M is endowed with a metric and connection, then the space of L_k^p sections ($k \in \mathbf{Z}^+, p \geq 1$) can be defined as a Banach space completion of $C^\infty(\xi)$ in the norm

$$
(6.31) \qquad \|\psi\|_{L_k^p} = \left(\int_M |\psi|^p + |\nabla\psi|^p + \cdots + |\nabla^k\psi|^p \right)^{1/p}
$$

Note that $L_\ell^2 = H_\ell$. The general *Sobolev Embedding Theorems* state

$$
(6.32) \qquad L_k^p \subset L_{k'}^{p'} \quad \text{if} \quad k - \frac{n}{p} \geq k' - \frac{n}{p'} \text{ and } k > k',
$$

where if $p' = \infty$ or $p = 1$, then we require strict inequality; and

$$
(6.33) \qquad L_k^p \subset C^\ell \quad \text{if} \quad k - \frac{n}{p} > \ell.
$$

The embedding (6.32) is compact if strict inequality holds, and therefore (6.33) is always compact. The *Sobolev Multiplication Theorems* state that

$$
(6.34) \qquad L_{k_1}^{p_1} \otimes L_{k_2}^{p_2} \to L_k^p
$$

is defined and continuous if

$$
\left(k_1 - \frac{n}{p_1} \right) + \left(k_2 - \frac{n}{p_2} \right) \geq \left(k - \frac{n}{p} \right),
$$

provided

$$k_1, k_2 \geq k \quad \text{and} \quad p_1 k_1, p_2 k_2 < n.$$

(Here $p_1, p_2 \neq 1$ and $p \neq \infty$.) The last restriction means among other things that $L_{k_i}^{p_i}$ in (6.34) do not consist of continuous functions (cf. (6.33)). In the continuous case $(pk > n)$, L_k^p is an algebra. Also in the continuous case, if $L_{k'}^{p'} \subset L_k^p$, then L_k^p is an $L_{k'}^{p'}$-module. Extreme caution is required in the borderline cases $p_i k_i = n$. The dual of L_k^p is the distribution space L_{-k}^q, where p and q are conjugate exponents $\left(\frac{1}{p} + \frac{1}{q} = 1\right)$. (Care must be taken if M has a boundary or is noncompact.) Multiplication theorems for these spaces follow by duality. Hence

$$L_{k_1}^{p_1} \otimes L_{-k}^q \to L_{-k_2}^{q_2}$$

is continuous iff (6.34) is, and dual to the continuous case $(pk > n)$ is the multiplication

$$L_k^p \otimes L_{-k}^q \to L_{-k}^q$$

and the fact that L_{-k}^q is an $L_{k'}^{p'}$-module if $L_{k'}^{p'} \subset L_k^p$. Composition on the right by a smooth function always maps L_k^p linearly to L_k^p. The (left) *Composition Lemma* states that composition on the left by a smooth function maps L_k^p to L_k^p in the Sobolev range $pk > n$.

The following *maximum principle* is easy to prove.

PROPOSITION 6.35. *Suppose $f \in C^2$ satisfies*

$$\Delta f + \gamma f \leq 0$$

on a compact domain $\mathcal{O}$, $\gamma \geq 0$, and $f \leq 0$ on $\partial \mathcal{O}$. Then $f \leq 0$ in $\mathcal{O}$.

We remind the reader that our sign convention on the Laplacian implies that $-\Delta f$ is the trace of the Hessian of f at a critical point. The next proposition is a *mean value inequality* [**GW**, §3], [**Mor**, Theorem 5.3.1].

PROPOSITION 6.36. *Suppose a positive function $f \in C^2$ satisfies $\Delta f \leq 0$ on a compact domain $\mathcal{O}$. Then for any $x \in \mathcal{O}$*

$$f(x) \leq c_{11} \int_{\mathcal{O}} f,$$

where c_{11} depends on $\mathrm{dist}(x, \partial \mathcal{O})$ and on the metric defining Δ.

Elliptic regularity states that a weak solution u of an elliptic system $Lu = 0$ is "smooth" in the interior of its domain. The precise definition of smooth depends on the smoothness of the coefficients of L. Since our metric is only C^k (§3), "smooth" for us does not really mean C^∞. Nevertheless, without loss of accuracy we can think of "smooth" as C^∞, and we will not be precise about this point. That said,

PROPOSITION 6.37. *If D is a self-dual connection, there is a (Sobolev) gauge transformation $s \in \mathcal{G}$ such that $s^*(D)$ is smooth.*

The proof (of an extended version) can be found in [**Pa**, §5], and we have more to say in §8 about this.

We have already used *unique continuation* [**Ar**] to prove that $\hat{\mathcal{M}}$ is a manifold (§3).

PROPOSITION 6.38. *Let L be a second order elliptic operator with scalar symbol defined on a connected domain, and assume that the coefficients of L are smooth. If $Lu = 0$ and u vanishes on an open set, then u vanishes identically.*

Unique continuation does not hold if the symbol of L is not a scalar.

Finally, we give the variational characterization of the smallest eigenvalue [**Tr**, §34].

PROPOSITION 6.39. *Let $\mathcal{D} : C^\infty(\xi) \to C^\infty(\xi')$ be a first order differential operator on a compact manifold, and suppose that $\mathcal{D}\mathcal{D}^* : C^\infty(\xi') \to C^\infty(\xi')$ is elliptic. Then if*

$$\alpha = \inf_{u \in C^\infty(\xi')} \frac{\int_M |\mathcal{D}^* u|^2}{\int_M |u|^2},$$

the equation

$$\mathcal{D}\mathcal{D}^* u - \alpha u = 0$$

has a smooth solution u.

Analytic Properties of SDYME.

The analysis in §7–§9 exploits the conformal invariance of the self-dual Yang–Mills equations. Let $g \to \kappa^2 g$ be a conformal change of metric. Then one-forms A, two-forms F, and the volume form vol transform according to

$$|A| \to \kappa^{-1}|A|$$

(6.40)
$$|F| \to \kappa^{-2}|F|$$

$$\mathrm{vol} \to \kappa^4 \, \mathrm{vol}.$$

The star operator is defined on two-forms by

$$F_1 \wedge {}^* F_2 = (F_1, F_2)\,\mathrm{vol},$$

and in view of (6.40) the right hand side is conformally invariant. The conformal invariance of the self-dual Yang–Mills equations (2.19) follows immediately. Also the norms

$$\|A\|_{L^4} = \left(\int_M |A|^4 * 1 \right)^{1/4}$$

$$\|F\|_{L^2} = \left(\int_M |F|^2 * 1 \right)^{1/2}$$

are conformally invariant, and they will be used to give conformally invariant estimates.

We quote two theorems of Karen Uhlenbeck and refer to the original papers for the proofs.

THEOREM 6.41 [**U1**, Theorem 1.5]. *Let D_n be a sequence of smooth connections, F_n the curvature of D_n, and suppose $\|F_n\|_{L^\infty} < c_{12}$ for all n. Then for a subsequence $\{n'\} \subseteq \{n\}$ there exist gauge transformations $s_{n'} \in \mathcal{G}$ and a connection D so that $s_{n'}^*(D_{n'}) \rightharpoonup D$ weakly in $L_1^p (1 < p < \infty)$ and $s_{n'}^*(D_{n'}) \to D$ strongly in L^p $(1 < p)$.*

Theorem 6.41 is essentially a compactness theorem—from a sequence of connections we produce a covergent subsequence. There is a more complete discussion in §8. The *Removable Singularities Theorem* states that connections with finite action have no point singularities. A proof is given in Appendix D.

THEOREM 6.42 [**U2**, Theorem 4.1]. *Fix $x \in M$, and suppose D is a self-dual connection on a bundle $\eta \to M \setminus \{x\}$ with finite action $\int_{M \setminus \{x\}} |F_D|^2 < \infty$. Then for some $s \in C^\infty(\mathrm{Aut}\,\eta)$, $s^*(\eta)$ extends to a smooth bundle $\overline{\eta} \to M$, and D extends to a smooth self-dual connection $\overline{D}$ on $\overline{\eta}$.*

Here $C^\infty(\mathrm{Aut}\,\eta)$ is the group of bundle automorphisms (gauge transformations) of η. There is no a priori control on $c_2(\overline{\eta})$, even if η is the restriction of a bundle over the entire manifold M. For self-dual connections on $SU(2)$ bundles, however,

$$c_2(\overline{\eta})[M] = -\frac{1}{8\pi^2} \int_M |F_D|^2 = -\frac{1}{8\pi^2} \int_{M \setminus \{x\}} |F_D|^2.$$

In this case $\overline{\eta}$ is completely described by the integer c_2, and hence the action $\int_{M \setminus \{x\}} |F_D|^2$ determines its topological type.

The grafting procedure of §6 provides us with a family of almost self-dual connections centered about any point $y \in M$. Now we want to perturb these to produce self-dual connections. Let D be an almost self-dual connection (we define this notion precisely later), and let F denote its curvature. Then the curvature F_A of a perturbation $D + A$ is

$$F_A = F + DA + A \wedge A,$$

whereby the anti-self-dual part of F_A is

$$P_- F_A = P_- F + P_- DA + P_- (A \wedge A).$$

Define $\mathcal{D} = P_- D$, and let

$$\# : \Omega^1(\mathrm{ad}\, \eta) \otimes \Omega^1(\mathrm{ad}\, \eta) \rightarrow \Omega^2_-(\mathrm{ad}\, \eta)$$

be the symmetric bilinear map

$$A_1 \# A_2 = \frac{1}{2} P_- (A_1 \wedge A_2 + A_2 \wedge A_1).$$

Then $P_- F_A = 0$ is the condition

$$(7.1) \qquad\qquad \mathcal{D} A + A \# A = -P_- F.$$

Unfortunately, (7.1) is not an elliptic equation—the gauge symmetry provides a kernel for the symbol and hence forbids ellipticity. We compensate by requiring A to have a special form which is essentially perpendicular to the gauge symmetry. The natural condition is to take $A = \mathcal{D}^* u$ for some $u \in \Omega^2_-(\mathrm{ad}\, \eta)$. Now

$$(7.2) \qquad\qquad \mathcal{L} u = \mathcal{D}\mathcal{D}^* u + \mathcal{D}^* u \# \mathcal{D}^* u = -P_- F$$

is a (nonlinear) elliptic equation, and if u satisfies (7.2), then $D + \mathcal{D}^* u$ is a self-dual connection. This chapter is devoted to the proof that solutions to (7.2) exist. Since $P_- F$ is small (D is almost self-dual), we expect that a solution $\mathcal{D}^* u$ will also be small, and the appropriate yardstick is the conformally invariant norm $\|_{L^4}$. This additional proviso (that $\|\mathcal{D}^*u\|_{L^4}$ be small) is included for technical reasons.

We solve (7.2) by the *continuity method*. Construct a curve $\mathcal{E}_t$ of equations

$$(7.3) \qquad \mathcal{E}_t : \mathcal{L}u_t = -tP_-F, \qquad t \in [0,1],$$

such that $\mathcal{E}_0$ is trivially solvable ($u_0 = 0$ is a solution) and $\mathcal{E}_1$ is the equation (7.2) we wish to solve. Set

$$I = \{t \in [0,1] : \mathcal{E}_t \text{ has a solution } u_t \text{ with } \|\mathcal{D}^*u_t\|_{L^4} \text{ small}\}.$$

(The auxiliary condition will be slightly modified later.) The continuity method entails proving that I is both open and closed. For then $I = [0,1]$ and $\mathcal{E}_1$ is solvable.

The openness of I follows if we prove that $\mathcal{L}$ is locally invertible about a solution u_t. Using the inverse function theorem, this is equivalent to the invertibility of the linear operator

$$(7.4) \qquad L_t(\Phi) = \mathcal{D}\mathcal{D}^*\Phi + 2\mathcal{D}^*u_t \,\#\, \mathcal{D}^*\Phi.$$

In particular, $L_0 = \mathcal{D}\mathcal{D}^*$ must be invertible; equivalently, the first eigenvalue α of $\mathcal{D}\mathcal{D}^*$ must be positive. We show this first. Here it is crucial that for λ sufficiently small, estimates on α be independent of λ. (λ is the parameter which measures the size of P_-F.) For λ is allowed to vary in an interval $(0,\overline{\lambda}]$, and $\overline{\lambda}$ is fixed only at the very end of the chapter. We remark that the positive definiteness of the intersection form enters at this stage. The invertibility of L_t follows easily from the invertibility of L_0.

The second step in the continuity method, the proof that I is closed, is in general more difficult. One first produces a priori bounds on solutions u_t. Then to each convergent sequence $t_n \to t_0$ in I corresponds a *weakly* convergent (sub)sequence $u_{t_n} \rightharpoonup u_{t_0}$. If the a priori estimates are strong enough, the limit u_{t_0} satisfies the equation $\mathcal{E}_{t_0}$. Success at this stage demands a creative choice of the space in which to solve $\mathcal{E}_t$—it must be strong enough to preserve the equation, yet weak enough so that a priori bounds can be proved.

We forged the analytic tools required for these arguments in §6. Recall the Weitzenböck formula (6.25) on 1-forms:

$$2\mathcal{D}^*\mathcal{D} + D D^* = \nabla^*\nabla + \mathrm{Ric}(\cdot) - 2[\cdot \lrcorner P_+F].$$

The curvature F_λ of our grafted connections blows up as $\lambda \to 0$—$\|F_\lambda\|_{L^\infty} = O(\lambda^{-2})$ by (6.12)—almost destroying the value of this formula for making estimates. Here is where we exploit conformal invariance to its utmost: we blow up the metric near y so that the curvature becomes bounded. However, our gain is balanced by a slight loss—now we must carry out our analysis on the *noncompact* manifold $M_y = M \backslash \{y\}$. Although noncompact manifolds are harder to handle, we are saved by the fact that near y the blown up manifold M_y is almost a uniform cylinder. This allows us to adopt standard techniques for compact manifolds by patching together results about the compact piece of M_y and the standard cylinder.

Taubes' projection is applied to the grafted connections to prove the existence of self-dual connections on M. We will find further applications to the Collar Theorem (§9), so that our exposition in this chapter is somewhat general. The original proof of Taubes' Theorem is in [**T**].

Blowing Up the Metric.

Consider the first flat metric δ on $\mathbf{R}^n$. Denoting the distance from the origin by r, form the conformally equivalent metric δ/r^2 on $\mathbf{R}^n \backslash \{0\}$. In polar coordinates $r, \theta^1, \ldots, \theta^{n-1}$ it is clear that $\langle \mathbf{R}^n \backslash \{0\}, \delta/r^2 \rangle$ is a cylinder distorted in the axial direction:

$$\delta/r^2 = \frac{dr^2}{r^2} + d\theta^2,$$

where $d\theta^2 = h_{ij} d\theta^i d\theta^j$ is the metric on the unit sphere S^{n-1}. The substitution $r = e^{-\tau}$ gives coordinates in which δ/r^2 is the standard product metric

$$\tilde{\delta} = d\tau^2 + d\theta^2$$

on the cylinder.

When we blow up an arbitrary metric g, we obtain a metric which is only approximately $\tilde{\delta}$. To make this precise we need a refinement of (6.13). Namely, the metric g can be written in geodesic polar coordinates as

$$g = dr^2 + g_{ij} d\theta^i d\theta^j,$$

where

$$g_{ij} = r^2 h_{ij} + O(r^4),$$

102

the error term involving the curvature of g [**CE**, §1.4]. Then

$$|g/r^2 - \delta/r^2| = O(r^2),$$

and in terms of τ,

$$(7.5) \qquad\qquad |\tilde{g} - \tilde{\delta}| = O(e^{-2\tau}).$$

Hence $\tilde{g}$ approaches the cylinder metric exponentially fast as $\tau \to \infty$. It follows that all derivatives of $\tilde{g}$ vanish exponentially fast as $\tau \to \infty$. For $\partial/\partial\theta^i$ this is clear. For $\partial/\partial\tau$, use $\partial/\partial\tau = -r\partial/\partial r$ and the fact that $g(r)$ is smooth in a neighborhood of the origin.

Apply this to a neighborhood of $y \in M$ to obtain a metric g_y on $M_y = M\backslash\{y\}$.

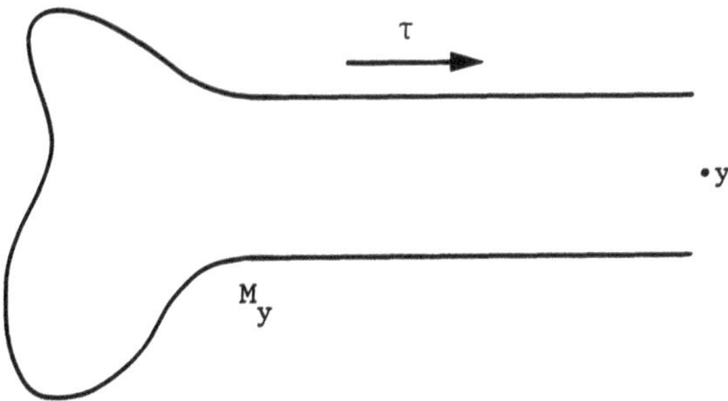

More precisely, let $\rho(M)$ be the minimum of injectivity radius of M and $3/4$. Choose a smooth function β satisfying

$$\beta(r) = \begin{cases} \dfrac{1}{r^2}, & 0 < r \le \frac{\rho(M)}{2}, \\ \text{decreasing}, & \frac{\rho(M)}{2} \le r \le \rho(M), \\ 1, & \rho(M) \le r. \end{cases}$$

Then the definition

$$g(x) \cdot \beta(\text{dist}(x,y))$$

makes sense in geodesic polar coordinates at y. Since g_y and its first two derivatives approach $\tilde{\delta}$ as $\tau \to \infty$, the curvature of g_y approaches (exponentially) the curvature of a cylinder as $\tau \to \infty$. So for any ε_1 (which we fix later), there exists $\tau_1(\varepsilon_1)$ such that

$$(7.6) \qquad\qquad |W^-_{g_y}(\tau,\theta)| < \varepsilon_1 \quad \text{if } \tau \ge \tau_1,$$

$$(7.7) \qquad\qquad |R(\tau,\theta) - 6| < \varepsilon_1 \quad \text{if } \tau \ge \tau_1,$$

since the cylinder is conformally flat ($W^+ = W^- = 0$) and has scalar curvature $R = 6$.

With the blown up metric at our disposal we can make precise the notion that a connection be *almost self-dual.* As in §6, B_r denotes the ball of radius r about y, and now let $C_{\bar\tau}$ denote the set where $\tau > \bar\tau$. Then a connection D is *λ-ASD (about y)* if its curvarture F satisfies

$$|F| \le c_{12} \quad \text{on } C_{\bar\tau}$$

$$|F| \le c_{13}\sqrt{\lambda} \quad \text{on } M_y \backslash C_{\bar\tau}$$

(7.8)
$$|P_-F| \le \varepsilon_2 \quad \text{on } C_{\bar\tau}$$

$$|P_-F| \le c_{14} \quad \text{on } M_y \backslash C_{\bar\tau}$$

$$\|P_-F\|_{L^2} \le c_{15}\lambda$$

for $\bar\tau = -\log(2\sqrt{\lambda})$. Here all norms refer to the metric g_y. (Remember, though, that $\|\,\|_{L^2}$ is conformally invariant on 2-forms.) The constants $c_{12} - c_{15}$ can be chosen arbitrarily, and ε_2 is to be determined later. The grafted connections D_λ are λ-ASD for $\lambda \le \lambda_1 = \min\left(\frac{\rho(M)^2}{4}, \frac{\varepsilon_2}{c_2}\right)$, since in the original metric g,

$$|F_\lambda| \le c_1\lambda^{-2} \quad \text{on } C_{\bar\tau}$$

$$|F_\lambda| = 0 \quad \text{on } M_y \backslash C_{\bar\tau}$$

$$|P_-F_\lambda| \le c_2 \quad \text{on } C_{\bar\tau}$$

$$|P_-F_\lambda| = 0 \quad \text{on } M_y \backslash C_{\bar\tau}$$

$$\|P_-F_\lambda\|_{L^2} \le c_2\lambda$$

by (6.12). Intuitively, a connection is λ-ASD if its curvature is almost self-dual and concentrated near y. The exact specifications are tailored to our needs.

We now prove that some standard results for compact manifolds carry over to M_y. Implicit in our notation henceforth is $\rho(M) \ge 1$, which we now assume for convenience.

PROPOSITION 7.9. *Let f be a twice differentiable function on M_y.*

    (i) *If $f, df \in L^1(M_y)$, then $\int_{M_y} \Delta f = 0$.*
    (ii) *$\|f\|_{L^4} \le c_{16}\|f\|_{H_1}$.*

Of course, (i) is integration by parts and (ii) is the Sobolev inequality (6.30).

PROOF: (i) Let $\beta_i \in C_0^\infty(M_y)$ be smooth compactly supported cutoff functions such that

$$\beta_i(\tau) = \begin{cases} 1 & \text{if } \tau \le i-1, \\ 0 & \text{if } \tau \ge i, \end{cases}$$

and $\beta_{i+1}(\tau) = \beta_i(\tau - 1)$.

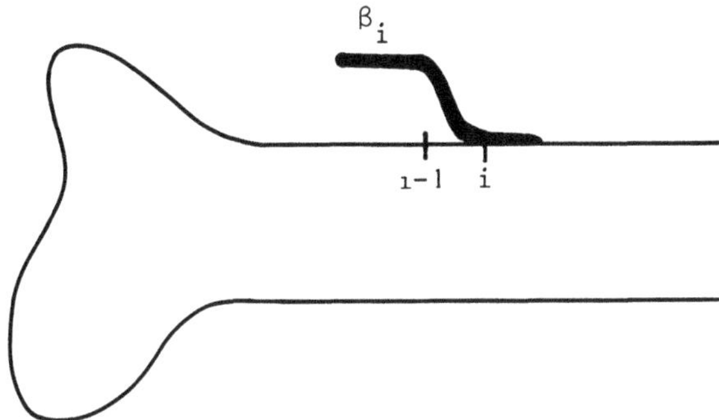

Then integration by parts yields

$$\left| \int_{M_y} \beta_i \Delta f \right| = \left| -\int_{M_y} f \Delta \beta_i + 2 \int_{M_y} (d\beta_i, df) \right|$$
$$\le \|\Delta \beta_i\|_{L^\infty} \int_{i-1 \le \tau \le i} |f| + 2\|d\beta_i\|_{L^\infty} \int_{i-1 \le \tau \le i} |df|$$

which tends to 0 as $i \to \infty$. The uniform estimates on derivatives of β depend on the almost uniformity of the metric (7.5).

(ii) Let c_{16} be the maximum of the Sobolev constant for M and for the cylinder. (Actually, c_{16} will be somewhat larger to account for the slight deviation of g_y from the cylinder metric.) Then if $f_i = f|_{i-1 \le \tau \le i}$, $f_0 = f$ restricted to

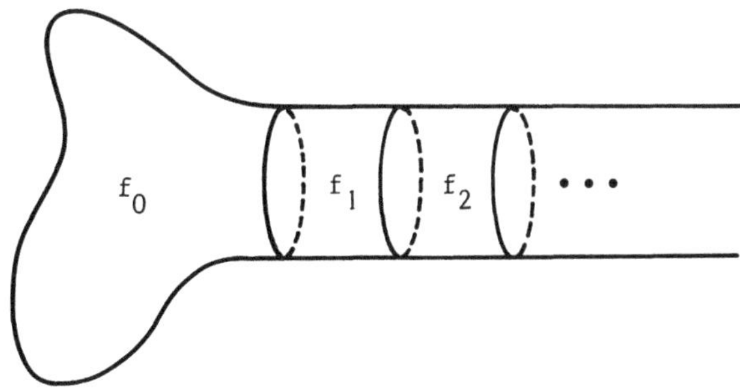

the compact part of M_y,

$$\left(\int_{M_y} |f|^4\right)^{1/2} \leq \sum_i \left(\int |f_i|^4\right)^{1/2}$$

$$\leq \sum_i c_{16} \int |f_i|^2 + |df_i|^2$$

$$= c_{16} \int_{M_y} |f|^2 + |df|^2$$

We state the following maximum principle in a form applicable both in the next section and in §9.

PROPOSITION 7.10. *Suppose $u(\tau, \theta)$ is a nonnegative function on the region $\{\tau_\ell \leq \tau \leq \tau_r\} \subset M_y$ satisfying*

  (i) $\Delta u + \gamma'^2 u \leq 0$,
  (ii) $u(\tau_\ell, \theta) \leq a_\ell$,
  (iii) $u(\tau_r, \theta) \leq a_r$.

Then

$$u(\tau, \theta) \leq a_\ell e^{-\gamma(\tau - \tau_\ell)} + a_r e^{-\gamma(\tau_r - \tau)}$$

for $\gamma = \gamma(\tau_\ell) < \gamma'$, with $\gamma(\tau_\ell) \to \gamma'$ as $\tau_\ell \to \infty$.

Again we remind the reader about our sign convention on the Laplacian (6.19).

PROOF: Note first that on $[t_\ell, t_r] \subset \mathbf{R}$, the function

$$v(t) = a_\ell e^{-\gamma(t - t_\ell)} + a_r e^{-\gamma(t_r - t)}$$

is a solution of the one dimensional Laplace equation

$$-\frac{d^2 v}{dt^2} + \gamma^2 v = 0$$

with boundary data $v(t_\ell) \geq a_\ell$, $v(t_r) \geq a_r$. Set $\tilde{v} = v \circ \tau$ on the "cylinder" $\{\tau_\ell \leq \tau \leq \tau_r\} \subset M_y$. Using the explicit form of the metric (7.5), we see that the standard formula for the Laplacian,

$$\Delta = -\frac{1}{\sqrt{g}} \frac{\partial}{\partial x^j} \left(\sqrt{g}\, g^{ij} \frac{\partial}{\partial x^i}\right)$$

simplifies, and compute

$$\Delta \tilde{v} = \left(-\frac{d^2 v}{dt^2} + O(e^{-2t})\frac{dv}{dt} \right) \circ \tau$$

$$\geq -(\gamma^2 + \gamma O(e^{-2\tau}))\tilde{v}$$

$$\geq -\gamma'^2 \tilde{v}$$

for appropriate γ. So

$$\Delta(u - \tilde{v}) + \gamma'^2(u - \tilde{v}) \leq 0,$$

and $u - \tilde{v} \leq 0$ at $\tau = \tau_\ell, \tau_r$. By the ordinary maximum principle (6.35), $u \leq \tilde{v}$ on $\{\tau_\ell \leq \tau \leq \tau_r\}$ as desired.

The Eigenvalue Estimate.

We prove that if D is λ-ASD for λ sufficiently small, then the first eigenvalue of $\mathcal{D}\mathcal{D}^*$ is positive. On noncompact manifolds elliptic operators may have a nondiscrete spectrum, and eigenfunctions do not exist in general. However, we can exercise our control over the noncompactness of M_y to demonstrate that the variational characterization of the first eigenvalue is still valid, at least if this eigenvalue is small.

LEMMA 7.11. *Suppose D is λ-ASD and define*

$$\alpha = \inf_{u \in \Omega^2_-(\operatorname{ad}\eta)} \frac{\int_{M_y} |\mathcal{D}^* u|^2}{\int_{M_y} |u|^2}.$$

Then if $\alpha < 1$, there exists $u \in L^2(M; \operatorname{ad}\eta \otimes \Lambda^2_- T^* M)$ satisfying the equation

$$\mathcal{D}\mathcal{D}^* u - \alpha u = 0.$$

Note that $L^2(\operatorname{ad}\eta \otimes \Lambda^2_- T^* M)$ can refer to either M or M_y, since the norm is conformally invariant on 4-manifolds.

PROOF: For large positive integers n define

$$\alpha_n = \inf_{\substack{u \in \Omega^2_-(\operatorname{ad}\eta) \\ u|_{\tau=n}=0}} \frac{\int_{M_y \setminus C_n} |\mathcal{D}^* u|^2}{\int_{M_y \setminus C_n} |u|^2}.$$

Then by (6.39) we can find $u_n \in L^2(M \backslash C_n)$ satisfying $\mathcal{D}\mathcal{D}^* u_n - \alpha_n u_n = 0$, and $\lim_{n \to \infty} \alpha_n = \alpha$ with α_n *decreasing*. (Remember the Trombone Principle (§3)!) Let $\tau_2 = \max(\tau_1, -ln\sqrt{2})$, and denote the compact set $M_y \backslash C_{\tau_2+1}$ by K. Unique continuation (6.38) shows that there is positive mass on K; hence we can normalize u_n so that

$$(7.12) \qquad \int_K |u_n|^2 = 1.$$

The Weitzenböck formula (6.26) gives the estimate

$$\int_K |\nabla u_n|^2 = \int_K (\nabla^* \nabla u_n, u_n)$$
$$= \int_K (2\mathcal{D}\mathcal{D}^* u_n + 2W^-(u_n) - \frac{R}{3} u_n - [P_- F, u_n], u_n)$$
$$\leq \left(2\alpha_n - \frac{R}{3} + 2\|W^-\|_{L^\infty(K)} + \|P_- F\|_{L^\infty(K)} \right),$$

which is bounded by (7.8). So $\{u_n\}$ is bounded in $H_1(K)$, and by Rellich's Theorem there is a strongly convergent subsequence $u_n \to \tilde{u}$ in $L^2(K)$. (We omit subsequence notation for convenience.)

Now work on the noncompact piece C_{τ_2}. Again we apply the Weitzenböck formula (6.26), this time aided by (6.18):

$$\frac{1}{2} \Delta |u_n|^2 \leq (\nabla^* \nabla u_n, u_n)$$
$$(7.13) \qquad \leq \left(2\alpha_n - \frac{R}{3} + 2\|W^-\|_{L^\infty(C_{\tau_2})} + \|P_- F\|_{L^\infty(C_{\tau_2})} \right) |u_n|^2$$
$$\leq (2\alpha_n - 2 + \frac{\varepsilon_1}{3} + 2\varepsilon_1 + \varepsilon_2)|u_n|^2$$

by (7.6), (7.7), and (7.8). For large n, we have $\alpha_n \sim \alpha$, and by choosing $\varepsilon_1, \varepsilon_2$ small and requiring α to be sufficiently small (the upper bound on α can be taken arbitrarily close to 1 as $\tau_2 \to \infty$), the expression in parentheses is negative. Then

$$(7.14) \qquad \Delta |u_n|^2 + \overline{\alpha} |u_n|^2 \leq 0 \text{ on } C_{\tau_2}, \ n \text{ large},$$

for some $\overline{\alpha} > 0$. By (7.14), (6.36), and (7.12) we can estimate

$$\max_{\tau = \tau_2} |u_n(\tau, \theta)|^2 \leq c_{11} \int_{\tau_2 - 1 \leq \tau \leq \tau_2 + 1} |u_n|^2 \leq c_{11} \int_K |u_n|^2 = c_{11}.$$

Recall that $|u_n|^2|_{\tau=n} = 0$. So applying (7.10) with $\tau_\ell = \tau_2$, $\tau_r = n$, we obtain

$$(7.15) \qquad |u_n(\tau,\theta)|^2 \le c_{11} e^{-\gamma(\tau-\tau_2)}$$

for some $\gamma < (\overline{\alpha})^{1/2}$. Integrating (7.15),

$$\int_{C_{\tau_2}} |u_n(\tau,\theta)|^2 \le \int_{C_{\tau_2}} c_{11} e^{-\gamma(\tau-\tau_2)} = c_{17}.$$

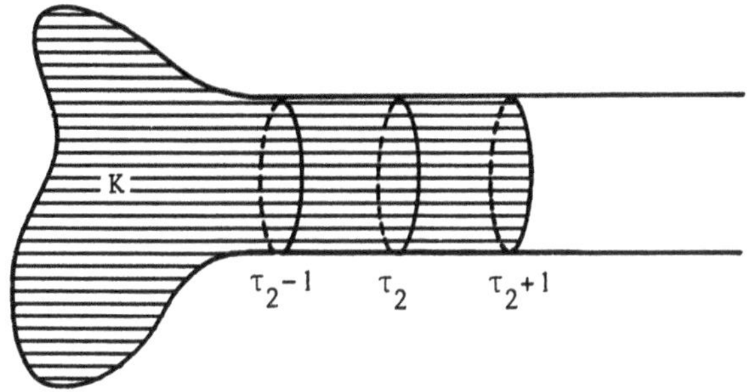

Hence we can extract a *weakly* convergent subsequence $u_n \rightharpoonup \overset{\approx}{u}$ in $L^2(C_{\tau_2})$. Of course, u_n still converges strongly to $\tilde{u}$ in K. Merge $\tilde{u}$ and $\overset{\approx}{u}$ to form u. Since linear equations are preserved under weak limits, $\mathcal{D}\mathcal{D}^* u - \alpha u = 0$ and $u \in L^2(M)$.

Now we invoke our topological assumptions to prove the eigenvalue estimate.

THEOREM 7.16. *There exists $\alpha > 0$ and $\lambda_2 > 0$ so that if D is λ-ASD about some $y \in M$, $\lambda \le \lambda_2$, then*

$$\alpha \int_{M_y} |u|^2 \le \int_{M_y} |\mathcal{D}^* u|^2.$$

PROOF: If not, fixing sequences $\lambda_n \to 0$, $\alpha_n \to 0$, we can find y_n, D_n, u_n such that D_n is λ_n-ASD about y_n and

$$\alpha_n \int_{M_{y_n}} |u_n|^2 = \int_{M_{y_n}} |\mathcal{D}_n^* u_n|^2.$$

By the lemma we can take u_n to be an eigenfunction of $\mathcal{D}_n \mathcal{D}_n^*$ with eigenvalue α_n. As M is compact, we may assume $y_n \to y$, and then M_{y_n} is approximately M_y as regards the metric. Equations (7.8) imply that the curvature $F_{D_n} \to 0$ on compact subsets K of M_y. Fixing K, it follows from the Compactness Theorem (6.41) and the absence of representations $\pi_1(M) \to SU(2)$ that up to gauge transformations a subsequence D_n converges to a trivial connection d on $\eta|_K$. Now repeat the argument of the lemma: normalize u_n so that $\int_K |u_n|^2 = 1$, and produce a weak limit $u \in L^2(M_y)$ satisfying $d_- d_-^* u = 0$. But $\eta|_{M_y}$ is a trivial bundle, and by taking a basis of flat sections we see that we have produced a nonzero harmonic anti-self-dual 2-form in $L^2(M_y) = L^2(M)$. This contradicts the positive definiteness of the intersection form ω.

We reiterate that our eigenvalue estimate is independent of $\lambda \leq \lambda_2$.

The Linearized Equation.

We now introduce the Hilbert space $\mathcal{H}$ on which we will solve equation (7.3):

$$\mathcal{L} u_t = \mathcal{D}\mathcal{D}^* u_t + \mathcal{D}^* u_t \ \# \ \mathcal{D}^* u_t = -t P_- F_A.$$

Our choice will depend on the connection D (and hence on λ). Define the Hilbert space $\mathcal{H}(D)$ as the completion of smooth compactly supported forms in $\Omega_-^2(\mathrm{ad}\,\eta)|_{M_y}$ with respect to the inner product

$$(7.17) \quad \{\Phi_1, \Phi_2\}_{\mathcal{H}(D)} = \int_{M_y} (\Phi_1, \Phi_2) + (\mathcal{D}^*\Phi_1, \mathcal{D}^*\Phi_2) + (\nabla\mathcal{D}^*\Phi_1, \nabla\mathcal{D}^*\Phi_2).$$

Intuitively, $\mathcal{H}(D)$ is a space of H_2 forms which, as illustrated by the following useful properties, has been adapted to our equation.

PROPOSITION 7.18.
   (i) $\mathcal{D}^* : \mathcal{H}(D) \to H_1(\mathrm{ad}\,\eta \otimes \Lambda_-^2 T^*M)$ is continuous.
   (ii) For $\Phi \in \mathcal{H}(D)$, $\|\mathcal{D}^*\Phi\|_{L^4} \leq c_{16}\|\Phi\|_{\mathcal{H}(D)}$.
   (iii) $\mathcal{L} : \mathcal{H}(D) \to L^2(\mathrm{ad}\,\eta \otimes \Lambda_-^2 T^*M)$ is smooth.

PROOF:
   (i) Let $\Phi \in \mathcal{H}$; it suffices to consider smooth compactly supported forms Φ. Then

$$\|\mathcal{D}^*\Phi\|_{H_1} = \left(\int |\mathcal{D}^*\Phi|^2 + |\nabla\mathcal{D}^*\Phi|^2 \right)^{1/2}$$

$$\leq \|\Phi\|_{\mathcal{H}(D)}.$$

(ii) This follows directly from (i) and the Sobolev inequality (7.9).

(iii) This follows from (i) and the fact that mult : $H_1 \otimes H_1 \to L^2$ is continuous (by the Sobolev Embedding Theorem and Hölder's inequality).

We study the linearized operator $L = \mathcal{D}\mathcal{D}^*$ at $t = 0$.

THEOREM 7.19. *There exists $c_{18} > 0$ such that if D is λ-ASD, $\lambda \leq \lambda_2$, then*

$$\|\Phi\|_{\mathcal{H}(D)} \leq c_{18}\|L\Phi\|_{L^2},$$

and $L : \mathcal{H}(D) \to L^2(\operatorname{ad}\eta \otimes \Lambda^2_- T^*M)$ is invertible.

PROOF: We prove invertibility by first demonstrating the inequality, which implies that L is injective with closed range, and then proving that the cokernel is empty.

Again we consider only $\Phi \in \mathcal{H}(D)$ which are smooth and compactly supported; our inequalities follow for all $\Phi \in \mathcal{H}(D)$ by a limiting argument. For smooth compactly supported forms, integration by parts is valid, and the eigenvalue estimate (7.16) yields

$$\begin{aligned}
\alpha \int_{M_y} |\Phi|^2 &\leq \int_{M_y} |\mathcal{D}^*\Phi|^2 \\
&= \int_{M_y} (\mathcal{D}\mathcal{D}^*\Phi, \Phi) \\
&\leq \left(\int_{M_y} |L\Phi|^2 \right)^{1/2} \left(\int_{M_y} |\Phi|^2 \right)^{1/2},
\end{aligned}$$

from which the estimates

$$\tag{7.20} \left(\int_{M_y} |\Phi|^2 \right)^{1/2} \leq \frac{1}{\alpha}\|L\Phi\|_{L^2}$$

$$\tag{7.21} \left(\int_{M_y} |\mathcal{D}^*\Phi|^2 \right)^{1/2} \leq \frac{1}{\sqrt{\alpha}}\|L\Phi\|_{L^2}$$

follow. To estimate $\int_{M_y} |\nabla\mathcal{D}^*\Phi|^2$ we apply the Weitzenböck formula (6.25) to $\mathcal{D}^*\Phi$:

(7.22)
$$\int_{M_y} |\nabla \mathcal{D}^*\Phi|^2 = \int_{M_y} (\nabla^*\nabla \mathcal{D}^*\Phi, \mathcal{D}^*\Phi)$$

$$= \int_{M_y} (2\mathcal{D}^*\mathcal{D}\mathcal{D}^*\Phi + \mathcal{D}\mathcal{D}^*\mathcal{D}^*\Phi - \operatorname{Ric}(\mathcal{D}^*\Phi) + 2[\mathcal{D}^*\Phi \lrcorner\, P_+F], \mathcal{D}^*$$

$$\leq 2\int_{M_y} |L\Phi|^2 + \int_{M_y} |D^*\mathcal{D}^*\Phi|^2$$

$$+ (\|\operatorname{Ric}\|_{L^\infty} + 2\|P_+F\|_{L^\infty}) \int_{M_y} |\mathcal{D}^*\Phi|^2.$$

Now $D^*\mathcal{D}^*\Phi = -[P_-F \lrcorner\, \Phi]$, so that

(7.23)
$$\int_{M_y} |D^*\mathcal{D}^*\Phi|^2 \leq \|P_-F\|_{L^\infty} \int_{M_y} |\Phi|^2.$$

The term $\|\operatorname{Ric}\|_{L^\infty}$ is bounded since it is essentially constant on the non-compact part of M_y, and $\|P_+F\|_{L^\infty}$, $\|P_-F\|_{L^\infty}$ are bounded by (7.8). (Here is where the blow-up pays off—in the g metric $\|P_+F\|_{L^\infty} = O(\lambda^{-2})$ and our estimate would fail.) Combining (7.20)–(7.23),

(7.24)
$$\left(\int_{M_y} |\nabla \mathcal{D}^*\Phi|^2\right)^{1/2} \leq c_{19}\|L\Phi\|_{L^2}.$$

Finally, (7.20), (7.21), and (7.24) imply the inequality of the theorem.

Suppose $\Phi \in \operatorname{Coker} L$. Let $\beta_i \in C_0^\infty(M_y)$ be a cutoff function as in (7.9). Then since integration by parts is valid for compactly supported forms,

$$0 = \int_{M_y} (\Phi, L(\beta_i^2 \Phi))$$

$$= \int_{M_y} (\mathcal{D}^*\Phi, \mathcal{D}^*(\beta_i^2 \Phi))$$

$$= \int_{M_y} \beta_i^2 |\mathcal{D}^*\Phi|^2 - \int_{M_y} 2\beta_i(\mathcal{D}^*\Phi, d\beta_i \lrcorner\, \Phi)$$

by (6.29). Therefore, Hölder's inequality implies

$$\int_{\operatorname{supp}(\beta_i)} \beta_i^2 |\mathcal{D}^*\Phi|^2$$

$$\leq 2\|d\beta_i\|_{L^\infty} \left(\int_{\operatorname{supp}(\beta_i)} \beta_i^2 |\mathcal{D}^*\Phi|^2\right)^{1/2} \left(\int_{\operatorname{supp}(d\beta_i)} |\Phi|^2\right)^{1/2},$$

form which

$$(7.25) \qquad \left(\int_{\mathrm{supp}(\beta_i)} \beta_i^2 |\mathcal{D}^*\Phi|^2 \right)^{1/2} \le 2\|d\beta_i\|_{L^\infty} \left(\int_{\mathrm{supp}(d\beta_i)} |\Phi|^2 \right)^{1/2}$$

But $\mathrm{supp}(d\beta_i) \subseteq \{i - 1 \le \tau \le i\}$ and $\Phi \in L^2$. Hence as $i \to \infty$ the right hand side of (7.25) tends to 0. Thus $\|\mathcal{D}^*\Phi\|_{L^2} = 0$, and from the eigenvalue estimate (7.16) we conclude that $\Phi = 0$. Therefore, $\mathrm{Coker}\, L = \{0\}$ and L is invertible.

Theorem 7.19 extends to include a lower order term of the type we will encounter.

THEOREM 7.26. *Suppose* $B \in L^4(\mathrm{ad}\,\eta \otimes T^*M)$, *and let* $L_B = L + B \,\#\, \mathcal{D}^*$. *Then there exists* $\delta > 0$ *such that if* $\|B\|_{L^4} < \delta$, *and* D *is* λ-*ASD,* $\lambda \le \lambda_2$, *then*

$$\|\Phi\|_{\mathcal{H}(D)} \le 2c_{18}\|L_B\Phi\|_{L^2},$$

and $L_B : \mathcal{H}(D) \to L^2(\mathrm{ad}\,\eta \otimes \Lambda_-^2 T^*M)$ *is invertible.*

In the application to the grafted connections (7.4), $B = 2\mathcal{D}^*u_t$ and $L_B\Phi = \mathcal{D}\mathcal{D}^*\Phi + 2\mathcal{D}^*u_t \,\#\, \mathcal{D}^*\Phi$.

PROOF: For compactly supported smooth Φ,

$$\begin{aligned}
\|\Phi\|_{\mathcal{H}(D)} &\le c_{18}\|L\Phi\|_{L^2} \\
&\le c_{18}(\|L_B\Phi\|_{L^2} + \|B \,\#\, \mathcal{D}^*\Phi\|_{L^2}) \\
&\le c_{18}(\|L_B\Phi\|_{L^2} + \|B\|_{L^4}\|\mathcal{D}^*\Phi\|_{L^4}) \\
&\le c_{18}(\|L_B\Phi\|_{L^2} + \delta c_{16}\|\Phi\|_{\mathcal{H}(D)})
\end{aligned}$$

by (7.18). Choose $\delta = \frac{1}{2c_{16}c_{18}}$ to obtain the desired inequality. L_B is injective with closed range as before.

To prove that L_B has no cokernel we use a continuity argument. The set $J = \{t \in [0,1] : L_{tB} \text{ is invertible}\}$ is nonempty and open (since invertibility is an open property). To prove that J is also closed, fix $\zeta \in L^2(\mathrm{ad}\,\eta \otimes \Lambda_-^2 T^*M)$, a sequence $t_n \to t_0$, and suppose $L_{t_n}\Phi_n = \zeta$. Then the inequality of the theorem gives a uniform bound $\|\Phi_n\|_{\mathcal{H}(D)} \le 2c_{18}\|\zeta\|_{L^2}$, whence we obtain a weak limit $\Phi_n \rightharpoonup \Phi_0$ by passing to a subsequence. Since linear equations are preserved under weak limits, $L_{t_0 B}\Phi_0 = \zeta$, $L_{t_0 B}$ is surjective. But $L_{t_0 B}$ is injective by the previous argument. Therefore, $L_{t_0 B}$ is invertible, J is closed, and hence L_B is invertible.

Taubes' Projection.

We finally have all of the ingredients necessary to produce self-dual connections.

THEOREM 7.27. *There exists $\lambda_3 > 0$ so that if D is λ-ASD, $\lambda \leq \lambda_3$, then for some $u \in \mathcal{H}(D)$, $D + \mathcal{D}^*u$ is a self-dual connection on M_y. Moreover u satisfies the estimate*

$$\|u\|_{\mathcal{H}(D)} \leq 2c_{18}\|P_- F_D\|_{L^2}.$$

Finally, $D + \mathcal{D}^*u$ extends to a connection on M.

PROOF: As described in the introduction to this chapter, we apply the continuity method to the equation

$$(7.28) \qquad \mathcal{E}_t : \mathcal{L}_t u_t = \mathcal{D}\mathcal{D}^* u_t + \mathcal{D}^* u_t \ \# \ \mathcal{D}^* u_t = -t P_- F_D$$

with the auxilliary condition

$$(7.29) \qquad \|u_t\|_{\mathcal{H}(D)} \leq c_{20}\lambda_3,$$

where λ_3 is to be specified and $c_{20} = 2c_{15}c_{18}$. Openness in t follows directly from Theorem 7.26 with $B = 2\mathcal{D}^* u_t$. The hypothesis of (7.26) is satisfied with $\lambda \leq \lambda_3$ where, since

$$\|2\mathcal{D}^* u_t\|_{L^4} \leq 2c_{16}\|u_t\|_{\mathcal{H}(D)} \leq 2c_{16}c_{20}\lambda_3,$$

we choose $\lambda_3 = \min(\lambda_2, \frac{\delta}{2c_{16}c_{20}})$.

To prove closedness in t we apply the a priori estimate of (7.26):

$$\|u_t\|_{\mathcal{H}(D)} \leq 2c_{18}\|-t P_- F_D\|_{L^2}$$

$$\leq 2c_{15}c_{18}\lambda_3 = c_{20}\lambda_3.$$

So if $t_n \to t_0$ and u_n satisfies the equation $\mathcal{E}_{t_n}$, we can extract a weakly convergent subsequence $u_n \rightharpoonup u_0$ in $\mathcal{H}(D)$. We must now check that the limit u_0 satisfies $\mathcal{E}_{t_0}$ and the auxilliary condition (7.29). First, since norms are lower semicontinuous in the weak topology,

$$\|u_0\|_{\mathcal{H}(D)} \leq \lim \|u_n\|_{\mathcal{H}(D)} \leq c_{20}\lambda_3,$$

so (7.29) checks. The linear term of (7.25) $\mathcal{D}\mathcal{D}^*u_n \to \mathcal{D}\mathcal{D}^*u_0$ is preserved under weak limits. For the nonlinear term we work on compact subsets; after all, we only have to show that the equation $\mathcal{E}_0$ is satisfied by u_0, and this is clearly a local question. Now the extended Sobolev Embedding Theorem (6.32) gives a strongly convergent subsequence $\mathcal{D}^*u_n \to \mathcal{D}^*u_0$ in $L^{4-\varepsilon}$ (using 7.18) for small $\varepsilon > 0$. Then by Hölder's inequality, $\mathcal{D}^*u_n \# \mathcal{D}^*u_n \to \mathcal{D}^*u_0 \# \mathcal{D}^*u_0$ in $L^{2-\frac{\varepsilon}{2}}$. So the equation is preserved in the limit.

The continuity method provides the desired solution u at $t = 1$. Elliptic regularity (6.37) guarantees that u is smooth in some gauge. Finally, the Removable Singularities Theorem (6.42) gives an extension to a self-dual connection on M. We must now show that the resulting connection lives on the $k = 1$ bundle (cf. the discussion following (6.42)). For any $\Phi \in \mathcal{H}(D)$, let $F(\Phi)$ denote the curvature of $D + \mathcal{D}^*\Phi$. It follows easily from (7.18) that the function $k : \mathcal{H}(D) \to \mathbf{Z}$ defined by

$$k(\Phi) = \frac{-1}{8\pi^2} \int_{M_y} tr(F(\Phi) \wedge F(\Phi))$$

is continuous. Thus $k(\Phi)$ is constant, and since $k(0) = 1$, we conclude that $D + \mathcal{D}^*u$ is a connection on the $k = 1$ bundle.

Combined with the grafting procedure of §6, Theorem 7.27 asserts the existence of nontrivial self-dual connections on M. This will lead to the construction of the collar in §9. The collar, a subset of $\mathcal{M}$, consists of *orbits* of self-dual connections, although we frequently choose a particular gauge and talk about self-dual connections. The local connectivity of this collar—two nearby self-dual connections with concentrated curvature can be joined by a curve of self-dual connections—will be important there, and as the proof is another application of our current circle of ideas, we present it here.

THEOREM 7.30. *Let D and D' be λ-ASD, $\lambda \le \lambda_3$. Suppose $D' = D + A$ with $\|A\|_{L^4} \le \frac{\delta}{4\sqrt{1+c_{16}c_{18}}}$. Then there exists a smooth family of self-dual connections*

$$D_t = D + tA + \mathcal{D}^*u_t$$

with $D_0 = D$ and $D_1 = D'$. Furthermore the curvatures $F_t = F_{D_t}$ satisfy

$$\|F_t - F\|_{L^2} \le t\|F' - F\|_{L^2} + \delta.$$

PROOF: We apply the continuity method to the equation

$$P_- F_{D_t} = 0$$

which, using the fact that $D + A$ is self-dual, can be written

$$(7.31) \qquad \mathcal{D}\mathcal{D}^* u_t + (2tA + \mathcal{D}^* u_t) \ \# \ \mathcal{D}^* u_t = (t - t^2)A \ \# \ A.$$

We also impose the auxilliary condition

$$(7.32) \qquad \|u_t\|_{\mathcal{H}(D)} \leq \frac{\delta}{2c_{16}}.$$

The proof proceeds as before, with $B = 2tA + \mathcal{D}^* u_t$ in Theorem 7.26. We have only to show that $u_1 = 0$. More generally, suppose that u_t, $\tilde{u}_t$ solve (7.31) subject to (7.32). Then

$$\mathcal{D}^* \mathcal{D}(u_t - \tilde{u}_t) + (2tA + \mathcal{D}^*(u_t + \tilde{u}_t)) \ \# \ \mathcal{D}^*(u_t - \tilde{u}_t) = 0,$$

and by (7.26), $\|u_t - \tilde{u}_t\|_{\mathcal{H}(D)} = 0$, so that $u_t = \tilde{u}_t$. Uniqueness at $t = 1$ gives $u_1 = 0$. The estimate on curvature follows easily from the expression

$$F_t = (1 - t)F_0 + tF_1 + (t^2 - t)(A \wedge A) + D\mathcal{D}^* u_t + (2tA - \mathcal{D}^* u)_t \wedge \mathcal{D}^* u_t.$$

The invariance of signature under oriented cobordism, a crucial ingredient in Donaldson's Theorem, depends on the compactness of the underlying manifolds. In this chapter we prove that $\overline{\mathcal{M}}$, the cobordism $M \sim \amalg^m \mathbf{CP}^2$ is compact.

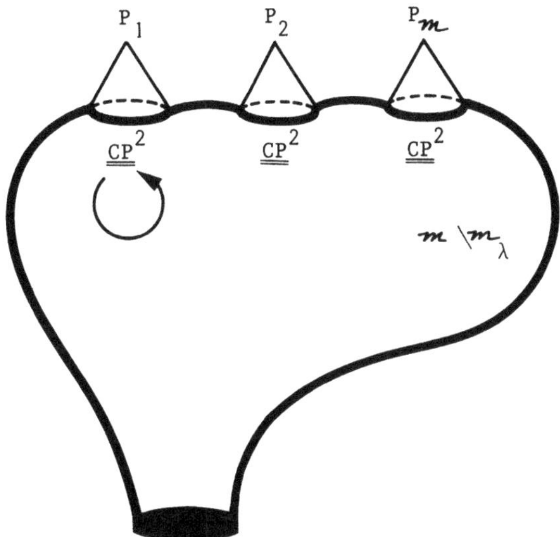

Actually, we only prove compactness for a suitably truncated version $\mathcal{M}\backslash\mathcal{M}_\lambda$ of our moduli space. In §9 we identify the end of $\mathcal{M}$ as $(0, \lambda) \times M$, and this allows us to deduce the compactness of $\overline{\mathcal{M}}$. Preliminary results needed in §9 will be derived in this chapter via compactness arguments.

Points of our moduli space are orbits of solutions to the self-dual Yang–Mills equations. Compactness theorems for solutions of nonlinear elliptic problems usually follow in a straightforward manner from the weak compactness of Sobolev spaces, but in our situation the gauge symmetry necessitates further arguments. We begin, then, by determining canonical local gauges for connections with small curvature. The existence of such Coulomb gauges immediately yields a regularity theorem for self-dual connections. Our argument here can be used quite generally to prove interior regularity for elliptic equations, so we give all of the details. Global compactness follows easily once we patch together the local results. Both the patching argument and the proof that Coulomb gauges exist can be found in [**U1**], and we do not repeat them here.

The collar region in $\mathcal{M}$ consists of orbits of connections whose curvature is concentrated in a ball near some point on M. Indeed, Taubes' construction produces connections with exactly this property. Now we must carry out the inverse procedure. Namely, given a connection with curvature concentrated near a point, we determine its center and scale. Here explicit formulas for instantons on S^4 (§6) play a role. We show that each concentrated instanton on M is "close" to one of these model instantons, and we measure its center and scale using this comparison. Only sufficiently concentrated connections allow measurement; these comprise the collar of $\mathcal{M}$. The chapter concludes with our main result: for any $\lambda > 0$, the subset of $\mathcal{M}$ where the scale is at least λ is compact.

Our arguments involve techniques from PDE, and again we explain the analysis carefully to benefit the novice.

Compactness and Regularity.

There are two types of compactness theorems in the literature. One implies that the set of connections $\{\overline{D} : \int |F_{\overline{D}}|^{2+\varepsilon} \leq B\}$ is weakly compact in $L_1^{2+\varepsilon}$ for any $\varepsilon > 0$ [**U1**]. (As always, the bar over a connection denotes its orbit under gauge transformations. Recall that the function $D \mapsto |F_D|$ is gauge invariant. Also, we continue to omit the volume form when no confusion is possible.) The topological type of the underlying bundle is preserved by weak $L_1^{2+\varepsilon}$ limits in four dimensions, since $L_1^{2+\varepsilon}$ functions are continuous. In his thesis [**Se**] Sedlacek proves that $\{\overline{D} : \int |F_{\overline{D}}|^2 \leq B\}$ is also "weakly" compact, but here the topology may change; the weak limit of $k = 1$ connections can have $k = 0$. Although neither result uses the self-dual equations, both strongly suggest that for a sequence of connections in our moduli space, either there is a covergent subsequence of the curvature concentrates near a point. This is our Theorem 8.36.

We begin with a local gauge fixing lemma. Let $\xi \to B_1$ be a (necessarily trivial) $SU(2)$ bundle over the four dimensional Euclidean unit ball. Any connection D on ξ can be written $D = d + A_\sigma$ once we choose a trivialization σ. Specify σ by requiring that the action

$$(8.1) \qquad \mathcal{E}(\sigma) = \int_{B_1} |A_\sigma|^2$$

be minimized. The Euler–Lagrange equation for $\mathcal{E}$ are

$$d^* A_\sigma = 0,$$

$$^* A_\sigma|_{S^3} = 0,$$

where the latter condition means that normal components of A_σ vanish on the boundary of the ball. Unfortunately, gauges constructed by direct minimization of (8.1) may have singularities. If we restrict to small curvatures, however, there are no such headaches.

LEMMA 8.2 [U1, Theorem 2.1]. *For some sufficiently small $\varepsilon_3 > 0$, if D satisfies $\|F_D\|_{L^2} < \varepsilon_3$, then there exists an L_2^2 gauge σ such that $D = d + A_\sigma$ with*

(i) $d^* A_\sigma = 0$,
(ii) $^* A_\sigma|_{S^3} = 0$,
(iii) $\|A_\sigma\|_{L_1^2} \le c_{21} \|F_D\|_{L^2}$.

Furthermore, σ and A_σ are unique up to constant gauge transformations.

Here L_2^2 norms of gauge transformations are computed via the inclusion $SU(2) \hookrightarrow \mathfrak{gl}(2)$, and L_k^p norms for sections of vector bundles are defined in (6.31). The proof proceeds by showing that $\{D : \|F_D\|_{L^2} < \varepsilon_3\}$ is connected, and that the set of D satisfying (i)–(iii) is both open and closed. Openness follows from the implicit function theorem (this fixes the value of ε_3), and closedness follows from compactness arguments as in §7. We refer to [U1] for the complete proof. Lemma 8.2 is valid on any size ball, since $\|F_D\|_{L^2}$ is conformally invariant.

For self-dual connections we can now deduce regularity.

PROPOSITION 8.3. *Let D be a connection on B_1, self-dual with respect to a metric g, and assume $\|F_D\|_{L^2} < \varepsilon_3$. Then there exists an L_2^2 gauge σ such that $D = d + A_\sigma$, A_σ is C^∞ in the half-sized ball $B_{1/2}$, and the estimate*

$$\|A_\sigma\|_{C^k(B_{1/2})} \le c_{22}(k) \|F_D\|_{L^2(B_1)}$$

holds. In particular,

$$\max_{|x| \le \frac{1}{2}} |F_D| \le c_{22}(1) \|F_D\|_{L^2(B_1)}.$$

Of course, we can replace $B_{1/2}$ by any $\mathcal{O} \subset\subset B_1$, but then $c_{22}(k)$ depends on $\rho = \text{dist}(\mathcal{O}, \partial B_1)$ and blows up as $\rho \to 0$. Also, the estimates are uniform over all metrics in a small C^k neighborhood of a fixed metric.

Local regularity for self-dual connections D on M follows immediately from (8.3). For since $\int_M |F_D|^2 < \infty$, every point x lies in a neighborhood $\mathcal{O}_x$ in which $\int_{\mathcal{O}_x} |F_D|^2 < \frac{\varepsilon_3}{2}$ and the metric is approximately constant. Blow up $\mathcal{O}_x$ so that it contains B_1. Then by conformal invariance of the L^2 norm on 2-forms, and since in normal coordinates the blown up manifold metric is conformally close to the Euclidean metric δ, the factor $1/2$ in $\varepsilon_3/2$ providing leeway to compensate for the difference, Lemma 8.2 applies. The resulting $A_{\sigma_x(\mathcal{O}_x)}$ is smooth by (8.3). Note, however, that the neighborhoods $\mathcal{O}_x$ are not uniform in size, so that the estimates in (8.3) are not uniform over M.

PROOF: Choose σ as in (8.2). Deleting the subscript σ for convenience, it follows from (8.2) and self-duality that $A \in L_1^2$ satisfies

$$d^* A = 0$$

$$d_- A + A \,\#\, A = 0.$$

Here $d_- = P_- d$, the anti-self-dual projection P_- taken with respect to g, whereas the adjoint d^* is taken with respect to δ. Thus A satisfies the single second order equation

$$(8.4) \qquad\qquad (dd^* + d_-^* d_-)A + d_-^*(A \,\#\, A) = 0,$$

where now the adjoint d_-^* is taken relative to g. The operator $L = dd^* + d_-^* d_-$ is elliptic, even though the adjoints refer to different metrics. Rather than tackle the known boundary conditions on A, we elect to use a compactly supported cutoff function $\varphi \in C_0^\infty(B_1)$ and deduce interior estimates. Employing the schematic notation ∇A for derivatives of A and $Q(\cdot, \cdot)$ for quadratic terms, (8.4) yields

$$(8.5)$$
$$(L+\mu)(\varphi A)+Q(A, \nabla(\varphi A)) = c_{23}(\varphi)A+c_{24}(\varphi)\nabla A+c_{25}(\varphi)Q(A, A)+\mu(\varphi A)$$

for any $\mu \in \mathbf{R}$. The functions $c_{23} - c_{25}$ also depend on derivatives of φ. Since L is a nonnegative elliptic operator, $L + \mu$ is invertible on Sobolev spaces for $\mu > 0$. More precisely, define $L_{1,0}^p$ to be the closure of C_0^∞ in L_1^p, and $L_{k,0}^p = L_k^p \cap L_{1,0}^p$ to be the subspace of L_k^p potentials which vanish on

∂B_1. (Elements in the closure of C_0^∞ in L_k^p also have vanishing derivatives at the boundary.) Then $L + \mu : L_{k,0}^p \to L_{k-2}^p$ is invertible. Furthermore, we can regard the first order term $Q(A, \nabla(\varphi A))$ in (8.5) as a bounded linear operator

$$Q(A, \nabla(\cdot)) : L_{1,0}^2 \to L_{-1}^2$$
$$: L_{2,0}^2 \to L^2$$

since the multiplications

$$L^2 \otimes L_1^2 \to L_{-1}^2$$
$$L_1^2 \otimes L_1^2 \to L^2$$

are continuous in four dimensions (6.34). Either way, its operator norm is bounded up to a constant by $\|A\|_{L_1^2}$, and by (8.2) this is small. Hence, readjusting ε_3 if necessary, this term is a small perturbation of $L + \mu$, and the operator $\tilde{L} = L + \mu + Q$ is invertible. Now the second multiplication rule above implies that the right hand side RHS of (8.5) is in L^2. So by the invertibility of $\tilde{L} : L_{2,0}^2 \to L^2$, there is a unique $\tilde{A} \in L_{2,0}^2$ satisfying

$$(8.6) \qquad\qquad\qquad \tilde{L}(\tilde{A}) = RHS.$$

But viewing $\tilde{A} \in L_{1,0}^2$ we see that (8.6) holds for $RHS \in L_{-1}^2$. Hence the invertibility of $\tilde{L} : L_{1,0}^2 \to L_{-1}^2$ and the fact that φA satisfies (8.5) imply $\varphi A = \tilde{A} \in L_{2,0}^2$, whence $A \in L_2^2(U)$ for any $U \subset\subset \operatorname{supp}\varphi$. Choose φ_2 with $\operatorname{supp}\varphi_2 \subset\subset \operatorname{supp}\varphi$, and rewrite (8.5) as

$$(8.7) \quad (L + \mu)(\varphi_2 A) = c_{23}(\varphi_2)A + c_{24}(\varphi_2)\nabla A + \varphi_2 Q(A, \nabla A) + \mu(\varphi_2 A).$$

Restrict to a domain U_2 satisfying $\operatorname{supp}\varphi_2 \subset\subset U_2 \subset\subset \operatorname{supp}\varphi$. We would like to say that the right hand side of (8.7) is in $L_1^2(U_2)$, but unfortunately while $L_2^2 \hookrightarrow L_1^4$, L_1^4 does not embed in L^∞, and so $L_2^2 \otimes L_1^2 \not\to L_1^2$. (This is the false borderline case of the Sobolev Embedding Theorem.) However, $Q(A, \nabla A) \in L_1^{2-\varepsilon}$ for any $\varepsilon > 0$, and we can apply the invertibility of $L + \mu : L_{3,0}^{2-\varepsilon} \to L_1^{2-\varepsilon}$ to deduce $A \in L_3^{2-\varepsilon}(U_2')$ for $U_2' \subset\subset \operatorname{supp}\varphi_2$. Now $Q(A, \nabla A) \in L_1^2(U_2')$ and we are off to the races: Iterate this procedure using equation (8.7) so that at the kth stage $A \in L_k^2(U_k)$. It is the continuous multiplication $L_k^2 \otimes L_{k-1}^2 \to L_{k-1}^2$, $k > 2$, which permits the bootstrapping. C^k estimates follows from the Sobolev embedding $L_{k+3}^2 \hookrightarrow C^k$. The constants $c_{22}(k)$ depend on $\|L + \mu\|^{-1}$ (hence on g), c_{21}, $\|\varphi_j\|_{C^j}(j \leq k)$, and ε_3, although the dependence on $\|\varphi_j\|_{C^j}$ can be used to express $c_{22}(k)$ in terms of $\mathrm{dist}(\mathcal{O}, \partial B_1)$ if we estimate on arbitrary $\mathcal{O} \subset\subset B_1$. Finally, the max estimate on $|F_D|$ is a consequence of the C^1 estimate on A.

From the regularity, the Arzela–Ascoli Theorem, and a patching argument, we deduce our global compactness result for compact domains $\Omega \subseteq M$.

THEOREM 8.8. *Let D_j be a sequence of connections on Ω, self-dual with respect to the metric g_j, and suppose $g_j \to g$ in $C^{k+1}(\Omega)$. If either*

(i) *there exists some $B < \infty$ with $\sup_{\Omega,j} |F_{\overline{D}_j}| \leq B$*

or

(ii) *for ε_3 as in Lemma 8.2, $\sup_{\Omega,j} \|F_{\overline{D}_j}\|_{L^2} < \frac{\varepsilon_3}{2}$,*

then for any $d > 0$ there exists a subsequence $\{j'\} \subseteq \{j\}$ and gauge transformations $s_{j'}$ so that $s_{j'}^*(D_{j'}) \to D$ in $C^k(\Omega_d)$ for some connection D which is self-dual relative to g.

Here $\Omega_d = \{x \in \Omega : \mathrm{dist}(x, \partial\Omega) \geq d\} \subset\subset \Omega$.

PROOF: Under either hypothesis we can cover Ω_d by a finite number of geodesic balls $\{B_\rho(x_\alpha)\}$ such that $B_{2\rho}(x_\alpha) \subset \Omega$, $\|F_{D_j}\|_{L^2(B_{2\rho}(x_\sigma))} < \varepsilon_3/2$, and the metric on $B_{2\rho}(x_\alpha)$ is C^{k+1} close to δ. Recalling that L^2 curvature estimates are dilation invariant, we can apply Proposition 8.3 to find L_2^2 gauge transformations $s_{j,\alpha}$ such that $s_{j,\alpha}^*(D_j) = d + A_{j,\alpha}$ in $B_{2\rho}(x_\alpha)$ with $A_{j,\alpha} \in C^{k+1}(B_\rho(x_\alpha))$. Since the metrics g_j converge in C^{k+1}, we have $\|A_{j,\alpha}\|_{C^{k+1}(B_\rho(x_\alpha))}$ is uniformly bounded for j large. Thus by the compactness of the inclusion $C^{k+1} \hookrightarrow C^k$, we can extract a subsequence $A_{j',\alpha}$ convergent in $C^k(B_\rho(x_\alpha))$ for each α. Now we must compare the various local gauges $s_{j',\alpha}$ and determine the limiting bundle. The success of this patching argument hinges on our control over the overlap functions

$$h_{j,\alpha\beta} = s_{j,\alpha} \circ s_{j,\beta}^{-1} : B_\rho(x_\alpha) \cap B_\rho(x_\beta) \to SU(2)$$

via the equation

$$dh_{j,\alpha\beta} = A_{j,\alpha} h_{j,\alpha\beta} - h_{j,\alpha\beta} A_{j,\beta}$$

as in (3.3). We refer to [**U1**, §3] for details, in particular for the fact that the topological type of the bundle is preserved in the limit. (The argument is somewhat simpler in our case since we have strong C^k convergence rather than weak Sobolev convergence.) We will illustrate some simple special cases of this patching in §9. Finally, the self-dual equations are preserved under strong limits, so the limiting connection is self-dual with respect to the limiting metric g.

After we learn how to measure concentrated curvature in the next section, we will apply Theorem 8.8 to deduce compactness results in $\mathcal{M}$.

Measuring Concentrated Curvature.

Recall from §6 that $k = 1$ instantons on S^4 are determined by a center $b \in S^4$ and a scale $\lambda > 0$. Geometrically, λ is the radius of the ball about b (in stereographic coordinates) which contains half the action:

$$(8.9) \qquad \int_{B_\lambda(b)} |F_{\lambda,b}|^2 = 4\pi^2.$$

Ultimately, we will compare self-dual connections on M whose curvature is concentrated with these instantons on S^4. Hence in this section we define a center and scale for such connections. Our construction is similar to that implicit in (8.9), except that we will integrate in normal coordinates and use a Friedrichs mollifier in place of the characteristic function of $B_\lambda(b)$.

We begin by studying nonnegative measures in a ball. These measures are thought of as the norm square curvature (field strength) of a connection on M multiplied by the volume form, or equivalently as $-tr(F \wedge *F)$. Although we restrict our attention to the ball B_2, our measures are defined in the larger ball B_4 in order that all of our definitions make sense. All our quantities depend smoothly on a C^k metric g, and measurements are made in that metric. Later we shall restrict g to be close to the Euclidean metric δ. Set

$$(8.10) \qquad \mathcal{F} = \{\omega \in L^1(B_4; \Lambda^4 \mathbf{R}^4) : \omega > 0, \int_{B_4} \omega < 8\pi^2, \int_{B_2} \omega > 4\pi^2\},$$

and let $\mathcal{C}_k$ denote the set of C^k metrics on B_4. For some smooth nonnegative cutoff function β satisfying

$$(8.11) \qquad \beta(\rho) = \begin{cases} 1 & \text{for } \rho < 1 - \delta, \\ 0 & \text{for } \rho \geq 1, \end{cases}$$
$$\beta'(\rho) \leq 0,$$

123

$\delta > 0$ a fixed small number, we define the smooth function

$$R : (0,2) \times B_2 \times \mathcal{F} \times \mathcal{C}_k \to \mathbf{R}$$

by

$$(8.12) \qquad R(\lambda, x, \omega, g) = \int_{B_4} \beta \left(\frac{\rho_g(x,z)}{\lambda} \right) \omega(z).$$

Here ρ_g is the distance function defined by the metric g. The cutoff function, which will be specified more precisely later, enables us to compute derivatives of R by differentiating β, and therefore properties of R will rely on the L^1 class of ω.

First, for each fixed $x \in B_2$ we can determine the radius $\lambda(x, \omega, g)$ of the ball centered at x which encompasses half of the total integral.

LEMMA 8.13. $\dfrac{\partial R}{\partial \lambda} > 0.$

PROOF: $\dfrac{\partial R}{\partial \lambda}(\lambda, x, \omega, g) = - \int_{B_4} \dfrac{\rho_g(x,z)}{\lambda^2} \beta' \left(\dfrac{\rho_g(x,z)}{\lambda} \right) \omega(z)$, and $\beta' \leq 0.$

Hence by the implicit function theorem, $\lambda(x, \omega, g)$ in

$$(8.14) \qquad R(\lambda(x, \omega, g), x, \omega, g) = 4\pi^2$$

is well-defined (our integral constraints guarantee that $4\pi^2$ is achieved for some $x \in B_2$). Note that $\omega \mapsto \lambda(x, \omega, g)$ is smooth. The scale (or width) of ω,

$$(8.15) \qquad \lambda(\omega, g) = \min_{x \in B_2} \lambda(x, \omega, g),$$

is now defined for all $\omega \in \mathcal{F}$, but in general the minimum is not attained at a unique point, so that the center cannot be defined. To obtain a unique minimum we must restrict to concentrated f's and metrics g close to δ. As a preliminary we examine our model: the basic instanton I in Euclidean space.

The field strength of I, which we view as a 4-form, is computed from (6.6) and (6.2) as

$$(8.16) \qquad \hat{I}(z) = \frac{48 dz}{(1 + |z|^2)^4}.$$

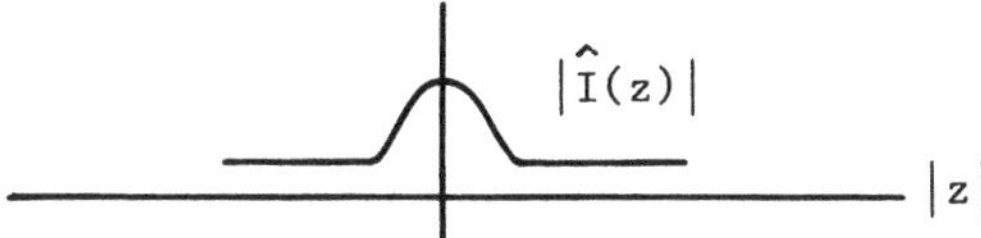

Eliminating the cutoff function and working in the metric δ, we set

$$(8.17) \qquad \hat{R}(\lambda, x) = \int_{B_\lambda(x)} \hat{I}(z),$$

and define $\hat{\lambda}(x)$ as in (8.14). Clearly $\hat{\lambda}$ achieves its minimum at 0, and it is easily computed that $\hat{\lambda}(0) = 1$. That there is a *unique* minimum is captured by the open properties in

LEMMA 8.18.

   (i) *The Hessian* $d_x^2 \hat{\lambda} > 0$ *in some ball* $B_{\varepsilon_4}(0)$.

   (ii) $\hat{\lambda} - 1 > c_{26} > 0$ *on* $B_2 \backslash B_{\varepsilon_4}$.

PROOF: The positivity of $d_x^2 \hat{\lambda}(0)$ is either obvious from the picture of $\hat{I}$, or is directly computable by implicity differentiating (8.14). (i) follows by continuity. Then $\hat{\lambda} > 1$ on $\partial B_{\varepsilon_4}$, so let $2c_{26} = \min_{\partial B_{\varepsilon_4}} \hat{\lambda} - 1$. Denoting the radial coordinate in $\mathbf{R}^4$ by r, we have $\frac{\partial \hat{R}}{\partial r} + \frac{\partial \hat{R}}{\partial \lambda} \frac{\partial \lambda}{\partial r} = 0$. But $\frac{\partial \hat{R}}{\partial \lambda} > 0$ by (8.13), and $\frac{\partial \hat{R}}{\partial r} < 0$ by direct calculation or from the following picture:

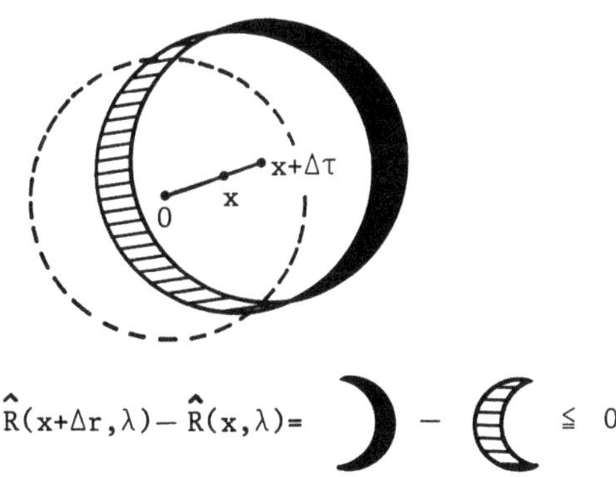

Hence (ii) holds.

We can rewrite (8.17) as

$$\hat{R}_{\hat{\beta}}(\lambda, x) = \int_{\mathbf{R}^4} \hat{\beta}\left(\frac{x-z}{\lambda}\right)\hat{I}(z)$$

(8.19)

$$= \lambda^4 \int_{\mathbf{R}^4} \hat{\beta}(z)\hat{I}(x+\lambda z)$$

for $\hat{\beta} = \chi_{|z|\leq 1}$, the characteristic function of the unit ball. Now $\hat{\beta}$ of the form $\hat{\beta}(z) = \beta(|z|)$, β satisfying (8.11), lie arbitrarily close to $\chi_{|z|\leq 1}$ in L^1, and we fix one such that $\hat{\lambda}_{\hat{\beta}}$ (obvious definition) satisfies (8.18) as well as

(iii)
$$\hat{\lambda}_{\hat{\beta}}(0) < 1 + c_{26}.$$

Notice that (i)–(iii) are open properties, and $\hat{\lambda}_{\hat{\beta}}$ is continuous in $\hat{\beta}$. With this choice of $\hat{\beta}$ we add a metric g to (8.19):

$$\hat{R}(\lambda, x, g) = \int_{\mathbf{R}^4} \beta\left(\frac{\rho_g(x,z)}{\lambda}\right)\hat{I}(z),$$

Again (i)–(iii) hold for $\hat{\lambda}(\cdot, g)$ if

(8.20)
$$\|g - \delta\|_{C^2} < \varepsilon_5$$

for some suitably small ε_5. Then $\hat{\lambda}(\cdot, g)$ has a unique minimum in B_{ε_4}.

Now we can specify which ω are centered.

PROPOSITION 8.21. *There exists $\varepsilon_6 > 0$ such that if $\|g - \delta\|_{C^2} < \varepsilon_5$ and $\omega \in \mathcal{F}$ satisfies $\|\omega - \hat{I}\|_{L^1(B_4)} < \varepsilon_6$, then $x(\omega, g)$ is a well-defined smooth function mapping into a small neighborhood of 0.*

PROOF: By implicit differentiation of (8.14) the Hessian of λ takes the form

$$d_x^2 \lambda(x, \omega, g) = \frac{\int_{B_4} f(\beta, \rho_g, x, z, \lambda)\omega(z)}{\int_{B_4} f_2(\beta, \rho_g, x, z, \lambda)\omega(z)},$$

where f also involves derivatives of β and ρ_g and $f_2 = \dfrac{\partial^2 f}{(\partial \lambda)^2}$. Hence (8.18)(i) is an L^1 open property of ω. (Recall that $\lambda(x, \omega, g)$ is continuous in ω.) Also, $\lambda(x, \omega, g) > 1 + c_{26}$ iff

$$\int_{B_4} \beta\left(\frac{\rho_g(x,z)}{1+c_{26}}\right)\omega(z) < 4\pi^2,$$

and this is an L^1 open statement. Similarly, $\lambda(0, \omega, g) < 1 + c_{26}$ iff

$$\int_{B_4} \beta \left(\frac{\rho_g(0, z)}{1 + c_{26}} \right) \omega(z) > 4\pi^2.$$

Choosing ε_6 appropriately, (i)–(iii) hold for $\lambda(x, \omega, g)$ replacing $\hat{\lambda}(x)$ and therefore $\lambda(\cdot, \omega, g)$ has a unique minimum.

So there is a well-defined map

$$\mathfrak{b} : \mathcal{F} \cap \{\omega : \|\omega - \hat{I}\|_{L^1(B_4)} < \varepsilon_6\} \times \{g \in \mathcal{C}_k : \|g - \delta\|_{C^2} < \varepsilon_5\}$$

(8.22)
$$\to (0, 2) \times B_2$$

$$\mathfrak{b}(\omega, g) = \langle \lambda(\omega, g), x(\omega, g) \rangle,$$

which is smooth for $k \geq 2$. Note that the 5-vector $\mathfrak{b} = \langle \lambda, x \rangle$ is the unique solution to the five implicit equations

$$R(\lambda, x, \omega, g) = 4\pi^2$$

(8.23)
$$\frac{\partial R}{\partial x}(\lambda, x, \omega, g) = 0$$

near $\langle \lambda, x \rangle = \langle 1, 0 \rangle$. Also, restricted to the neighborhood specified in (8.22),

(8.24)
$$\|d_\omega \mathfrak{b}\| \leq c_{27},$$

(8.25)
$$\|d_\omega \mathfrak{b}(\omega, g_1) - d_\omega \mathfrak{b}(\omega, g_2)\| = O(\|g_1 - g_2\|_{C^2}),$$

since $R(\lambda, x, \omega, g)$ depends linearly on ω and smoothly on the C^2 class of g. Here, as always, $\mathfrak{b}(\cdot, g)$ depends on the L^1 class of ω, and so the norms in (8.24) and (8.25) are operator norms relative to the L^1 norm in the domain.

We want to extend our construction to measures close to the model *concentrated* instantons

(8.26)
$$\hat{I}_\lambda(z) = \frac{48\lambda^4 dz}{(\lambda^2 + |z|^2)^4}$$

(cf. (6.8), (6.2)). To transfer data from M to our local balls, for each $y \in M$, $\lambda > 0$ we define

(8.27)
$$e_{\lambda, y} = \exp_y \circ T_\lambda,$$

where $T_\lambda : \mathbf{R}^4 \to \mathbf{R}^4$ is dilation by λ, and we identify $T_y M \simeq \mathbf{R}^4$. Although this identification can only be made up to an element of $SO(4)$, our geometric data—the forms $\hat{I}_\lambda$ and transformations T_λ—are $SO(4)$ invariant. Now our measures ω and metrics g are defined directly on M. Assume for convenience that the injectivity radius of M is at least 4.

THEOREM 8.28. *There exists $\lambda_4 > 0$ such that if $\omega \in L^1(M; \Lambda^4 T^* M)$ satisfies*

$$\|e^*_{\mu,y}(\omega) - \hat{I}\|_{L^1(B_4)} < \varepsilon_6$$

for some $\mu \leq \lambda_4$, $y \in M$, and

$$\int_M \omega \leq 9\pi^2 \quad (= 8\pi^2 +),$$

then for the functional

$$R(\lambda, x, \omega, g) = \int_M \beta\left(\frac{\rho_g(x,z)}{\lambda}\right) \omega(z),$$

the equations

$$R(\lambda, x, \omega, g) = 4\pi^2$$
$$\frac{\partial R}{\partial x}(\lambda, x, \omega, g) = 0$$

have a unique solution

$$\mathfrak{b}(\omega, g) = \langle \lambda(\omega, g), x(\omega, g) \rangle \in (0, 2\lambda_4) \times M$$

which depends smoothly on ω and g. Moreover,

$$\|d_\omega \mathfrak{b}\| = O(\lambda)$$
$$\|d_\omega \mathfrak{b}(\omega, g) - d_\omega \mathfrak{b}(\omega, \delta)\| = O(\lambda^3),$$

where $\lambda = \lambda(\omega, g)$, and δ is the flat metric in exponential coordinates at $x(\omega, g)$.

PROOF: Set $\tilde{\omega} = e^*_{\mu,y}(\omega)$ and $\tilde{g} = e^*_{\mu,y}(g)/\mu^2$ on $B_4 \subset T_y M$. Then $\|\tilde{\omega} - \hat{I}\|_{L^1(B_4)} < \varepsilon_6$ by assumption, and $\|\tilde{g} - \delta\|_{C^2} = O(\mu^2)$ which, for λ_4 sufficiently small, is less than ε_5. Hence the definition

$$\mathfrak{b}(\omega, g) = \langle \mu \cdot \lambda(\tilde{\omega}, \tilde{g}), \; e_{\mu,y}(x(\tilde{\omega}, \tilde{g})) \rangle$$

makes sense by (8.21). Furthermore, since we have pulled back the metric, this definition is independent of the choice of μ and of y in a small ball, and the integral constraint guarantees that y can be chosen in this ball. Since $\lambda(\tilde{\omega}, \tilde{g}) \leq 2$, we have $\lambda(\omega, g) \leq 2\lambda_4$, so that the estimate (8.24) yields

$$\|d_\omega \mathfrak{b}(\omega, g)\| = \mu \|d_\omega \mathfrak{b}(\tilde{\omega}, \tilde{g})\| = 0(\lambda),$$

and by (8.25),

$$\|d_\omega \mathfrak{b}(\omega, g) - d_\omega(\omega, \delta)\| = \mu \|d_\omega \mathfrak{b}(\tilde\omega, \tilde g) - d_\omega \mathfrak{b}(\tilde\omega, \tilde\delta)\|$$
$$= O(\lambda^3),$$

where $\lambda = \lambda(\omega, g)$.

We are finally ready to define a center $\lambda(D)$ and a scale $x(D)$ for concentrated connections D. Given such a connection we define the 4-form

$$(8.29) \qquad \omega(D) = -tr F_D \wedge *F_D = |F_D|^2 * 1.$$

Note that $\omega(D)$ is gauge invariant.

DEFINITION. $\mathcal{CC} = \{D : \text{for some } \mu \le \lambda_4 \text{ and } y \in M, \text{ the inequalities}$ $\|e^*_{\mu,y}(\omega(D)) - \hat I\|_{L^1} < \varepsilon_6 \text{ and } \int_M \omega(D) \le 9\pi^2 \text{ hold}\}$.

Endow $\mathcal{CC}$ ("concentrated connections") with the L_1^2 topology, i.e., if $D, D' \in \mathcal{CC}$ and $D' = D + A$, then $\|D - D'\| = \|A\|_{L_1^2}$. Define

$$\mathcal{B}(D) \;=\; \mathfrak{b}(\omega(D), g)$$

for a fixed metric g on M. Our results in this section are summarized in

THEOREM 8.30. *The map*

$$\mathcal{B} : \mathcal{CC} \to (0, \lambda_4) \times M$$

is smooth and satisfies

$$|d\mathcal{B}_D(A)| \le c_{28} \lambda(D) \|DA\|_{L^2(M)}.$$

PROOF: By the chain rule,

$$d\mathcal{B}_D = d\mathfrak{b}_{\omega(D)} \circ d\omega_D.$$

Differentiate (8.29) to obtain

$$d\omega_D(A) = -2tr F_D \wedge *DA.$$

So using the estimate in (8.28) and Cauchy–Schwartz,

$$|d\mathcal{B}_D(A)| = O(\lambda) \, \|tr F_D \wedge *DA\|_{L^1},$$
$$= O(\lambda) \, \|F_D\|_{L^2} \, \|DA\|_{L^2},$$

from which the theorem follows.

By the gauge invariance of our constructions, $\mathcal{B}$ factors through the orbit space

$$
\begin{array}{ccc}
\mathcal{SD} \cap \mathcal{CC} & & \\
 & \searrow \mathcal{B} & \\
\downarrow & & (0, \lambda_4) \times M. \\
 & \nearrow \overline{\mathcal{B}} & \\
\mathcal{M}_{\lambda_4} \subseteq \mathcal{M} \cap \mathcal{CC} & &
\end{array}
$$

Here we have set $\mathcal{M}_\lambda = \{\overline{D} \in \mathcal{M} : \lambda(\overline{D}) < \lambda\}$. In §9 we prove that for λ sufficiently small, $\overline{\mathcal{B}} : \mathcal{M}_\lambda \to (0, \lambda) \times M$ is a diffeomorphism.

Compactness in $\mathcal{M}$.

Although we have precisely defined a class $\mathcal{CC}$ of concentrated connections, we have not yet shown that $\mathcal{CC}$ is nonempty. In fact, connections whose curvature blows up lie in $\mathcal{CC}$, and these were constructed in §7. The following theorem, then, is a sort of inverse to Taubes' grafting procedure.

THEOREM 8.31. *Suppose $\{\overline{D}_j\}$ is a sequence in $\mathcal{M}$ with $\max_M |F_{\overline{D}_j}| \to \infty$. Then for any $K > 0$ there exist lifts D_j of $\overline{D}_j$, points $x_j \in M$, and numbers $\lambda_j < \frac{1}{K}$ such that*

   (i) $\lambda_j \to 0$;
   (ii) $e_{\lambda_j, x_j}(D_j) \to I_K$ in $C^k(B_K)$.

Here I_K is the basic instanton whose concentration is given by (8.16), but restricted to the ball $B_K \subset T_{x_j} M \approx \mathbf{R}^4$, and $e_{\lambda, x}$ is defined in (8.27).

PROOF: Set

$$
\max_M |F_{\overline{D}_j}| = |F_{\overline{D}_j}|(x_j) = \frac{1}{\lambda_j^2} \to \infty.
$$

Notice that x_j may not be uniquely determined (simply pick one if there are several), and λ_j, x_j are *not* $\lambda(\overline{D}_j)$, $x(\overline{D}_j)$ as defined in the last section. For each j, fix a lift D_j of $\overline{D}_j$. The connection $\tilde{D}_j = e^*_{\lambda_j, x_j}(D_j)$ is self-dual with respect to $g_j = e^*_{\lambda_j, x_j}(g)$, g the metric on M, and since the self-dual equations are conformally invariant, $\tilde{D}_j$ is also self-dual with respect to $\tilde{g}_j = g_j/\lambda_j^2$. It follows easily from (6.13) that $\tilde{g}_j \to \delta$, uniformly on the larger ball B_{K+1}. Since $\max_{B_K} |F_{\tilde{D}_j}| = 1$, we can apply the Global Compactness Theorem (8.8) to conclude that for a subsequence $\{j'\} \subset \{j\}$, and for possibly different lifts $D_{j'}$, we have $\tilde{D}_{j'} \to D_K$ in $C^k(B_K)$ with the limiting connection D_K self-dual in the metric δ. Using a diagonal process we can arrange that a subsequence converge simultaneously on all B_K, $K \in \mathbf{Z}$, and thus obtain a limiting connection D on $\mathbf{R}^4$ with $\max_{\mathbf{R}^4} |F_D| = |F_D|(0) = 1$. By the Removable Singularities Theorem (6.42), D extends to a self-dual connection on S^4. Furthermore,

$$0 < \int_{\mathbf{R}^4} |F_D|^2 = \lim_{K \to \infty} \lim_{j' \to \infty} \int_{B_K} |F_{\tilde{D}_{j'}}|^2 \leq 8\pi^2,$$

so that D is a $k = 1$ instanton (cf. the discussion following (6.42)). The classification of $k = 1$ instantons on S^4 (§6) shows that D is gauge equivalent to the basic instanton. Finally, an amusing lemma in metric space theory—if a sequence in a metric space X has the property that every subsequence has in turn a subsubsequence which converges to $p \in X$, then the original sequence converges to p—obviates the need to take subsequences when we restrict to a fixed compact set; i.e., $\tilde{D}_j \to I_K$ in $C^k(B_K)$. This follows since any subsequence of $\{\overline{D}_j\}$ satisfies the hypotheses of the thoerem, and the argument above produces a subsubsequence converging in $C^k(B_K)$ to I_K. However, we cannot conclude that $\tilde{D}_j \to I$ in $C^K(\mathbf{R}^4)$ since we have no uniform estimates on the rate of convergence as K varies.

Under the hypotheses of the theorem we can deduce the following corollaries, of which only the first is immediately relevant for compactness. Corollary 8.35, which describes concentrated instantons away from their centers, will be useful in §9.

COROLLARY 8.32. $\lambda(\overline{D}_{j'}) \to 0$.

PROOF: Recall that $\lambda(\overline{D})$ is defined for any $\overline{D} \in \mathcal{M}$ and is continuous in $\overline{D}$. Since $\tilde{D}_{j'} \to I_K$, $\lambda(I_K) = 1$, and $\lambda(\overline{D}_{j'}) = \lambda_{j'} \cdot \lambda(\tilde{D}_{j'})$, the assertion follows.

Using Taubes' construction we can find $\{\overline{D}_j\}$ satisfying the hypotheses of (8.31), and $\overline{D}_j \in \mathcal{CC}$ for large j by (8.32). Hence the sets $\mathcal{M}_\lambda$ of concentrated instantons are nonempty.

The next corollary is essentially a restatement of (8.31).

COROLLARY 8.33. Suppose $\{\overline{D}_j\}$ is a sequence in $\mathcal{M}$ with $x = x(\overline{D}_j)$ constant and $\lambda_j = \lambda(\overline{D}_j) \to 0$. Then $e^*_{\lambda_j, x}(D_j) \to I_K$ in $C^k(B_K)$ for some lifts D_j.

PROOF: By the definition of $\lambda(\overline{D}_j)$,

$$\int_M \beta \left(\frac{\rho_g(x, z)}{\lambda_j} \right) |F_{\overline{D}_j}|^2(z) * 1 = 4\pi^2,$$

and since $\lambda_j \to 0$, it follows easily that $\max_M |F_{\overline{D}_j}| \to \infty$. Now apply (8.31).

Reverting now to the hypotheses of (8.31), we assume (by passing to a subsequence if necessary) that $x_j \to x$.

COROLLARY 8.34. *For any $\rho > 0$,*

$$\lim_{j \to \infty} \int_{M \setminus B_\rho(x)} |F_{\overline{D}_j}|^2 = 0.$$

PROOF: Since $\lambda_j \to 0$, $B_{\lambda_j}(x) \subset B_\rho(x)$ for large j, and since $\tilde{D}_j \to I_{1/\rho}$ in $B_{1/\rho}$, we have $\int_{B_\rho(x)} |F_{\overline{D}_j}|^2 \to 8\pi^2$.

COROLLARY 8.35. *The sequence $\{D_j|_{M \setminus B_\rho(x)}\}$ is gauge equivalent to a sequence which converges to a trivial connection on $\eta|_{M \setminus B_\rho(x)}$.*

PROOF: Apply the Global Compactness Theorem (8.8) on $\Omega = M \setminus B_\rho(x)$. Because the resulting limit D is a flat connection (by (8.34) its curvature vanishes), and since there are no nontrivial representations $\pi_1(\Omega) = \pi_1(M) \to SU(2)$, D is trivial. Here again we do not have to take a subsequence since only one limit is possible.

Finally, we prove that the truncated moduli space is compact.

THEOREM 8.36. *$\mathcal{M} \setminus \mathcal{M}_\lambda$ is compact for any $\lambda > 0$.*

PROOF: Let $\{\overline{D}_j\}$ be a sequence in $\mathcal{M} \setminus \mathcal{M}_\lambda$. Since $\lambda(\overline{D}_j) \nrightarrow 0$, $\max_M |F_{\overline{D}_j}|$ is uniformly bounded by (8.31) and (8.32). Now our Global Compactness Theorem (8.8) produces a convergent subsequence of lifts $D_{j'} \to D$ in $C^k(M)$, and hence $\overline{D}_{j'} \to \overline{D}$ in $\mathcal{M} \setminus \mathcal{M}_\lambda$.

We complete the proof of Donaldson's Theorem in this chapter by showing that for λ sufficiently small,

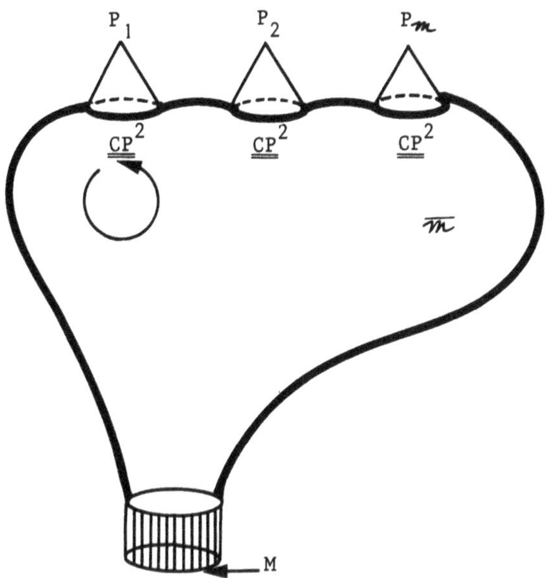

$\mathcal{M}_\lambda = \{\overline{D} \in \mathcal{M} : \lambda(\overline{D}) \leq \lambda\}$ is diffeomorphic to $(0, \lambda) \times M$. Recall from (8.30) that for $\lambda \leq \lambda_4$ there is a well-defined smooth map

$$\overline{B} : \mathcal{M}_\lambda \to (0, \lambda) \times M$$
$$\overline{D} \mapsto \langle \lambda(\overline{D}), x(\overline{D}) \rangle.$$

Here $x(\overline{D})$ is the center of the instanton $\overline{D}$ and $\lambda(\overline{D})$ is its width.

THEOREM 9.1. *There exists $\lambda_8 > 0$ such that for $\lambda \leq \lambda_8$, the map $\overline{B}$ is a diffeomorphism.*

The proof of this Collar Theorem proceeds in several steps. First, $\overline{B}$ is proper; this is merely a restatement of the Compactness Theorem (8.36).

Next, $\overline{B}$ is a local diffeomorphism; we outline the argument. Recall that for our model space S^4, the conformal group acts transitively on the moduli space. The conformal transformation on $\mathbf{R}^4$ consisting of translation by b and dilation by μ transforms the basic instanton into one whose center is b and scale is μ. Infinitesimally, this gives a 5 parameter family of deformations on which $d\overline{B}_{S^4}$ acts as the identity. When transferred to our more general situation, these deformations are not quite conformal, and hence do not preserve the self-dual equations. Nevertheless, with the aid of decay estimates on curvature, we show that the deviation from self-duality is small, and so an infinitesimal Taubes projection (7.19) straightens out these deformations. Again we have a 5 parameter space on which $d\overline{B}$ is invertible, completing the proof that $\overline{B}$ is a local diffeomorphism. Proper local diffeomorphisms are finite covering maps, whence $\overline{B}$ is a global diffeomorphism if it is 1:1. Suppose $\overline{B}(\overline{D}) = \overline{B}(\overline{D}')$. Then a suitable choice of gauge yields lifts D, D' of $\overline{D}, \overline{D}'$ for which $\|D - D'\|_{L^4}$ is small. By our local connectivity result (7.30), D and D' are joined by a short path in $\mathcal{M}$, and this path cannot leave the collar since the distance between a concentrated instanton and an unconcentrated instanton is large. Therefore, $\overline{D} = \overline{D}'$, and the proof of (9.1) is complete.

Our presentation follows Donaldson's original proof, although we have added many details and made a few technical modifications.

Decay Estimates.

By now we have three different local models for concentrated instantons D_λ: the standard model I of scale λ, the normalized model II of scale 1, and the blown-up, cylinder model III. We will use all three interchangeably, and for convenience we display them here along with their metrics.

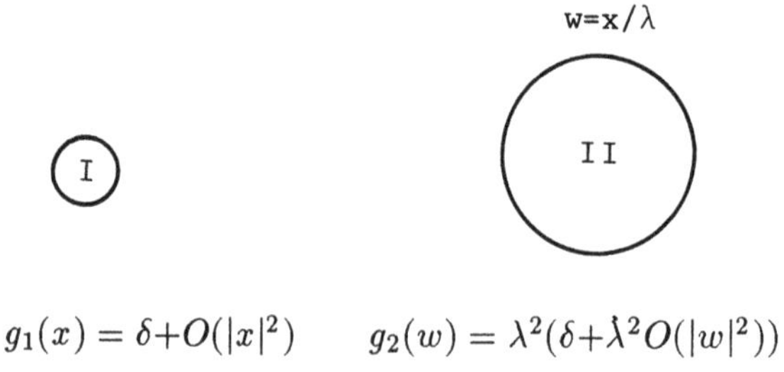

$$g_1(x) = \delta + O(|x|^2) \qquad g_2(w) = \lambda^2(\delta + \lambda^2 O(|w|^2))$$

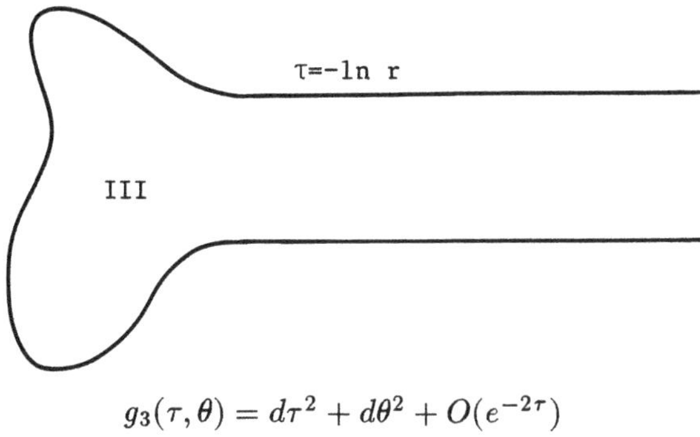

$$g_3(\tau, \theta) = d\tau^2 + d\theta^2 + O(e^{-2\tau})$$

Divide the blown-up instanton into three overlapping regions: a compact piece $M_0 = \{\tau \leq \tau_0 + 1\}$, a "cylinder" $I_\lambda = \{\tau \geq \tau_\lambda - 1\}$, and a "neck" $N_\lambda = \{\tau_0 \leq \tau \leq \tau_\ell\}$ in between.

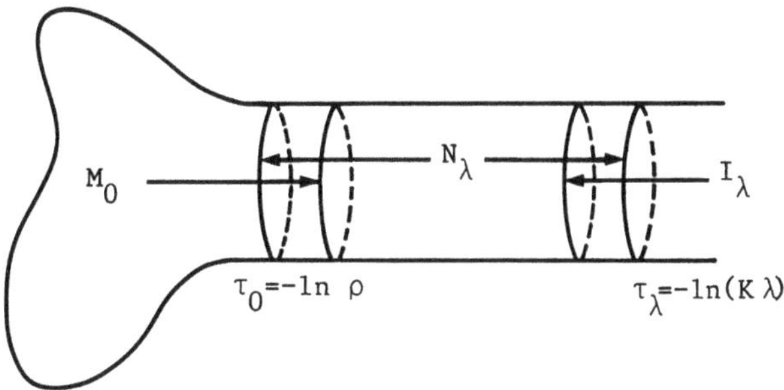

(The neck of the instanton is not to be confused with the collar of the moduli space!) Here ρ is a small constant fixed by the curvature of M near $x(D_\lambda)$, and $\tau_0 = -ln\rho$ plays the role of τ_ℓ in (7.10). The right endpoint $\tau_\lambda = -ln(K\lambda)$; the large constant K will be fixed only at the very end of the argument. The scale λ takes values in $(0, \overline{\lambda}]$, where $\overline{\lambda}$ depends on K and is always chosen so that $K\overline{\lambda}$ remains small relative to ρ.

We summarize the results from §8 as follows: D_λ looks like a standard instanton on I_λ, and D_λ is almost flat on the compact piece M_0, both statements improving as $\lambda \to 0$. However, we have as yet no precise control of D_λ in the neck region $N_\lambda = \{K\lambda \leq r \leq \rho\}$. To determine what sort of decay to expect, we revert to our model instantons. The curvature (cf. (6.8))

(9.2)
$$|F_\lambda|(r, \theta) = \frac{\sqrt{48}\lambda^2}{(\lambda^2 + r^2)^2}$$

satisfies

$$(9.3) \qquad |F_\lambda|(r,\theta) = \frac{c_{29}K^2\lambda^2}{r^4}, \qquad r \geq K\lambda.$$

The difference between (9.3) and our final estimate (9.8) reflects the fact that the discarded term in (6.21), $|\nabla F| - |d|\,|F|$, is of the same order as the other terms. We note that the precise value of the coefficient of R in the Weitzenböck formula (6.26) plays a role in this section (cf. Appendix C). As this formula is rather confusing to compute (indeed, several previously published versions are incorrect), we are comforted by the existence of another proof of (9.8) with an even better decay estimate [U2], [D]. These other proofs rely on the exact eigenvalue of a certain operator on S^3.

Corollary 8.34 gives a rough L^2 estimate on curvature away from the center of D_λ. The following max estimate is much more precise.

PROPOSITION 9.4. *We have*

$$\lim_{\lambda \to 0} \max_{\tau \leq \tau_\lambda} |F_{D_\lambda}| \leq \frac{\nu}{2},$$

where

$$\nu = \frac{c_{30}}{K^2}.$$

Here and throughout this section the norm refers to the blown-up metric g_3. The factor of 2 is for convenience later.

PROOF: The Regularity Theorem (8.3) implies that

$$|F_{D_\lambda}|^2(\tau,\theta) \leq c_{31}(r) \int_{B_r(\tau,\theta)} |F_{D_\lambda}|^2.$$

Since M_0 is compact and the geometry of N_λ is locally uniform, we can choose balls B_r of uniform size covering $M_0 \cup N_\lambda$. Hence

$$(9.5) \qquad \max_{\tau \leq \tau_\lambda} |F_{D_\lambda}|^2 \leq c_{31} \int_{\tau \leq \tau_\lambda + 1} |F_{D_\lambda}|^2.$$

By conformal invariance, the explicit formula (9.2), and the convergence (8.33) to the standard instanton,

$$\lim_{\lambda \to 0} \int_{\tau \geq \tau_\lambda + 1} |F_{D_\lambda}|^2 = 8\pi^2 - \int_{|w| \leq \frac{K}{e}} \frac{48 dw}{(1 + |w|^2)^4},$$

$$(9.6) \qquad\qquad\qquad \leq \frac{24\pi^2}{\left(1 + K^2/e^2\right)^2}.$$

The desired result follows from (9.5) and (9.6).

Next, we exploit the minimizing property of self-dual connections to show that the curvature on N_λ bounds the curvature on M_0.

PROPOSITION 9.7. *Suppose D is a self-dual connection satisfying*

$$|F_D|(\tau, \theta) \leq \varepsilon_7, \qquad \tau_m - 1 \leq \tau \leq \tau_m,$$

for some $\varepsilon_7 < 1$ and $\tau_m \geq \tau_0$. Then

$$|F_D|(\tau, \theta) \leq c_{32}\varepsilon_7, \qquad \tau \leq \tau_m,$$

for some $c_{32} > 0$ not depending on D.

The conclusion is also valid on M_0.

PROOF: In an exponential gauge (cf. (9.38)) we can write $D = d + A$ on $\tau_m - 1 \leq \tau \leq \tau_m$ with

$$|A| \leq c_{33}\varepsilon_7.$$

Let φ be a cutoff function near τ_m,

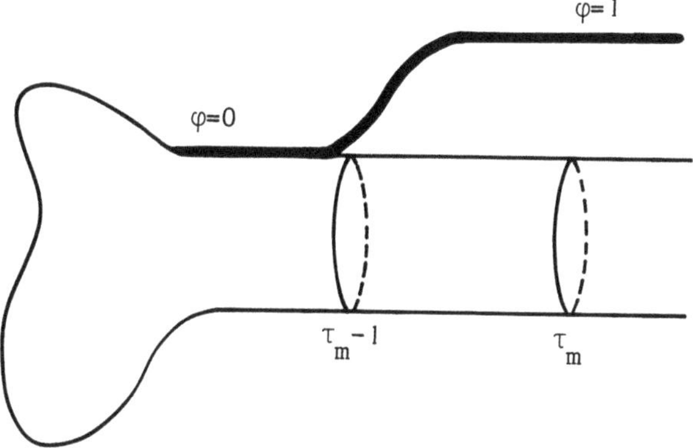

and form a new, non-self-dual connection

$$D_\varphi = \begin{cases} D & \text{on } \tau \geq \tau_m, \\ d + \varphi A & \text{on } \tau_m \geq \tau \geq \tau_m - 1, \\ \text{flat} & \text{on } \tau_{m-1} \geq \tau. \end{cases}$$

Then since D is an absolute minimum for $\mathcal{YM}$ (2.13),

$$\int_{\tau \geq \tau_m} |F_D|^2 + \int_{\tau \leq \tau_m} |F_D|^2 = \int_M |F_D|^2$$

$$\leq \int_M |F_{D_\varphi}|^2$$

$$\leq \int_{\tau \geq \tau_m} |F_D|^2 + c_{34}\varepsilon_7.$$

The max estimate now follows from regularity as in the previous proposition.

We proceed to our decay estimate. Let D_λ be a self-dual connection of scale λ, and denote its curvature by F_λ.

THEOREM 9.8. *Given any $\gamma < \sqrt{2}$, we can choose $K \geq K_1 = K_1(\gamma)$ and ρ sufficiently small so that for $\lambda \leq \lambda_5 = \lambda_5(\rho, K)$,*

$$|F_\lambda|(\tau, \theta) \leq c_{35} \nu e^{-\gamma(\tau_\lambda - \tau)}, \qquad \tau_0 \leq \tau \leq \tau_\lambda,$$

relative to the blown-up metric g_3. This translates into

$$|F_\lambda|(x) \leq c_{35} \nu \left(\frac{K\lambda}{|x|}\right)^\gamma \left(\frac{1}{|x|}\right)^2, \qquad K\lambda \leq |x| \leq \rho$$

relative to the manifold metric g_1.

PROOF: By reversing the orientation of the underlying manifold in (6.26), we derive a Weitzenböck formula for self-dual 2-forms:

$$(9.9) \qquad 2D_+ D_+^* = \nabla^* \nabla - 2W^+(\cdot) + \frac{R}{3} + [P_+ F, \cdot] \text{ on } \Omega_+^2(\mathrm{ad}\,\eta),$$

where $D_+ = P_+ D$. As in (7.13) we obtain for $F = F_\lambda$

$$\Delta |F| + 2|F| \leq (2\varepsilon_1 + \frac{\varepsilon_1}{3} + |F|)|F|$$

on the "cylinder" $\tau_0 \leq \tau \leq \tau_\lambda$. (Instead of (6.18), use the more delicate (6.21) to make this estimate.) Then $\varepsilon_1 \to 0$ as $\rho \to 0$ by (7.6) and (7.7), and $|F| \to 0$ as $K \to \infty$ by (9.4). Hence

$$\Delta |F| + \gamma' |F| \leq 0, \qquad \tau_0 \leq \tau \leq \tau_\lambda,$$

and $\gamma' < 2$ can be taken arbitrarily close to 2. For small λ we also have $|F|(\tau_\lambda, \theta) \leq \nu$ by (9.4), and we temporarily denote

$$a = \max_\theta |F|(\tau_0, \theta).$$

The maximum principle (7.10) now yields the estimate

$$(9.10) \qquad |F|(\tau, \theta) \leq a e^{-\gamma(\tau - \tau_0)} + \nu e^{-\gamma(\tau_\lambda - \tau)}, \qquad \tau_0 \leq \tau \leq \tau_\lambda,$$

for $\gamma < \sqrt{\gamma'}$. Recall that γ can be taken arbitrarily close to $\sqrt{\gamma'}$ if we let $\tau_0 \to \infty$ (i.e., $\rho \to 0$). By elementary calculus the function in (9.10) achieves its minimum value at

$$\tau_m = \frac{1}{2}(\tau_0 + \tau_\lambda) + \frac{ln(a/\nu)}{2\gamma},$$

and the minimum value is exactly

$$2\sqrt{a\nu}e^{-\gamma(\tau_\lambda - \tau_0)/2}.$$

It follows from (9.7) that we can take a to be on the order of this minimum:

$$a \leq 2c_{32}\sqrt{a\nu}e^{-\gamma(\tau_\lambda - \tau_0)/2},$$

or, by squaring,

$$a \leq 4c_{32}^2\nu e^{-\gamma(\tau_\lambda - \tau_0)}.$$

Feeding back into (9.10), we have

$$(9.11) \qquad |F|(\tau,\theta) \leq \nu e^{-\gamma(\tau_\lambda - \tau)}\left[1 + 4c_{32}^2 e^{-2\gamma(\tau - \tau_0)}\right],$$

from which the desired estimate is immediate.

We remark that the terms thrown away in (9.11) and in the maximum principle (7.10) are insignificant.

Donaldson's proof of the decay estimates is quite different from ours, as he relies on a relative Chern class formula for manifolds with boundary. His proof works for self-dual fields, whereas our estimate is valid for any minimizing Yang–Mills field. (These are not necessarily the same—consider, for example, manifolds with boundary.) Donaldson points out that the Removable Singularities Theorem (7.42) can be proved from these estimates. We include the argument in Appendix D.

Conformal Deformations.

For each sufficiently concentrated self-dual connection $D \in \mathcal{CC}$, we exhibit a 5 parameter family of deformations $\Sigma = \Sigma_{\mu,b} : \mathbf{R}^5 \to T_D\mathcal{SD}$ such that $d\mathcal{B} \circ \Sigma$ is onto. Since $\mathcal{M}$ is 5-dimensional, and $\mathcal{B}(D) = \overline{\mathcal{B}}(\overline{D})$ factors through the moduli space, it follows that $d\overline{\mathcal{B}}$ is invertible, which proves that $\overline{\mathcal{B}}$ is a local diffeomorphism. Recall that $\mathcal{B}$ is defined implicitly by a functional

$R(\lambda, x, \omega, g)$ which depends smoothly on $\omega = |F_{\overline{D}}|^2 * 1$ and on C^k metrics g. In the flat case,

$$R(\lambda, x, \omega) = \int_{\mathbf{R}^4} \beta \left(\frac{|x - z|}{\lambda} \right) \omega(z)$$

for an appropriate cutoff function β, and a simple change of variable yields

$$R(\lambda, x, \omega) = R(\mu\lambda, \mu x + b, T^*_{\mu,b}(\omega)),$$

where $T_{\mu,b}(y) = \mu y + b$ is the conformal transformation consisting of dilation by μ and translation by b. It follows that the map

$$(9.12) \qquad \langle \mu, b \rangle \mapsto \langle \lambda(T^*_{\mu,b}(\omega)), x(T^*_{\mu,b}(\omega)) \rangle = \langle \mu\lambda(\omega), \mu x(\omega) + b \rangle$$

is a diffeomorphism of a neighborhood of $\langle 1, 0 \rangle$ onto a neighborhood of $\langle \lambda, x \rangle \in \mathbf{R}^5$. In particular, if $x(\omega) = 0$ (which we can always arrange in an appropriate coordinate system), its differential at 0 is the identity map if $\lambda = 1$, and is

$$\left(\begin{array}{c|c} \lambda & 0 \\ \hline 0 & id \end{array} \right)$$

in general. To take care of the factor of λ, we define the variations

$$(9.13) \qquad \Sigma_{\mu,b} = \frac{d}{dt}\Big|_{t=0} T^*_{1+t\frac{\mu}{\lambda}, tb}(\omega);$$

then

$$\langle \mu, b \rangle \mapsto d_\omega \mathbf{b} \circ \Sigma_{\mu,b}$$

is the identity. Note that the action in (9.13) is simply the Lie derivative by the vector fields

$$(9.14) \qquad \sum_k \left(b^k + \frac{\mu}{\lambda} x^k \right) \frac{\partial}{\partial x^k}.$$

These vector fields, as well as the conformal transformations they generate, lift to act on connections (§6), and setting

$$\omega(D) = -tr\, (F_D \wedge *F_D) = |F_D|^2 * 1,$$

we see that the conformal invariance of $*$ on 2-forms implies

$$T^*_{\frac{\mu}{\lambda},b}(\omega(D)) = \omega \left(T^*_{\frac{\mu}{\lambda},b}(D) \right).$$

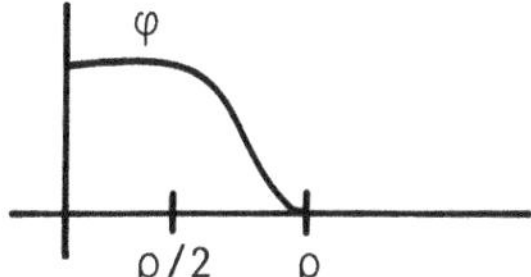

Moreover, if D is self-dual, then $T^*_{\mu,b}(D)$ is also, since the Yang–Mills equations are conformally invariant. So the differential of the composite map

$$\langle \mu, b \rangle \mapsto T^*_{\frac{\mu}{\lambda},b}(D) \overset{B}{\mapsto} \langle \lambda(\cdot), x(\cdot) \rangle$$

is exactly the identity map in the flat case.

In our curved situation the infinitesimal conformal transformations (9.14) can be transferred to M via a cutoff function φ. Fix once and for all a self-dual connection D with $\lambda(D) = \lambda$. We define the 5 parameter family of vector fields

$$(9.15) \qquad X_{\mu,b} = \varphi(|x|) \sum_k \left(b^k + \frac{\mu}{\lambda} x^k \right) \frac{\partial}{\partial x^k}$$

in normal coordinates on M near $x(D)$. Intuitively, $X_{\mu,b}$ is infinitesimal dilation by $\frac{\mu}{\lambda}$ and translation by b near $x(D)$, which is cut off in a ball of radius ρ. We will lift to an action on connections below, but there are two main differences from the flat case. First, the manifold metric g_1 is not flat, so the vector fields (9.15) no longer induce the identity on dB. However, g_1 is almost flat near the origin, so that the deviation from the identity is small. The second difficulty is that $X_{\mu,b}$ is not conformal, and so does not preserve the self-dual equations; i.e., the lift to the space of connections is not tangent to the self-dual connections $\mathcal{SD}$. But near the origin $X_{\mu,b}$ is nearly conformal, and away from the origin the curvature decays rapidly. Hence the anti-self-dual curvature picked up is small. Now an infinitesimal Taubes projection (7.19) can be applied, and the total correction is small. Thus we obtain a 5 parameter family

$$(9.16) \qquad \Sigma_{\mu,b} = L_{X_{\mu,b}} D + \mathcal{D}^* \Phi_{\mu,b}$$

of infinitesimal variations of D tangent to $\mathcal{SD}$. This outlines the proof of

THEOREM 9.17. *For* $\lambda(D) \le \lambda_6$,

$$d\mathcal{B}_D \circ \Sigma : \mathbf{R}^5 \to \mathbf{R}^5$$

is invertible.

Since $\mathcal{B}$ is constant along orbits of the gauge action, the five dimensional image of Σ in $T_D\mathcal{SD}$ projects onto $T_{\overline{D}}\mathcal{M}$, proving

COROLLARY 9.18. *For* $\lambda \le \lambda_6$,

$$\overline{\mathcal{B}} : \mathcal{M}_\lambda \to (0,\lambda) \times M$$

is a local diffeomorphism.

We begin by defining the action of $X = X_{\mu,b}$ on connections. In general, diffeomorphisms on manifolds do not lift to bundles. However, lifts of the generators always exist, and any two differ by an infinitesimal automorphism of the bundle covering the infinitesimal identity map on the base, i.e., by an infinitesimal gauge transformation. Since we eventually mod out by gauge transformations, and in any case are only interested in norms of curvature, which are independent of gauge, the particular lift we choose is irrelevant. Just to fix a convention, lift X to the horizontal vector field $\tilde{X}$ on the principal bundle P, where "horizontal" is defined by the connection D. Then the lift acts on connections, in particular on D, and we denote the action by

$$(9.19) \qquad\qquad L_X D = [\tilde{X}, D].$$

The induced action on curvature is

$$(9.20) \qquad\qquad L_X F_D = [\tilde{X}, F_D] = D(L_X D).$$

Of course, the action on the metric is the usual Lie derivative. The Lie derivative notation in (9.19), (9.20) is appropriate from several points of view. For example, the various actions are computed by Lie differentiating the base variables and ignoring the fiber variable. Either the bracket notation, or explicit local formulas like

$$L_{a^k \frac{\partial}{\partial x^k}}(g_{ij}dx^i \otimes dx^j) = \left(a^k \frac{\partial g_{ij}}{\partial x^k} + g_{kj}\frac{\partial a^k}{\partial x^i} + g_{ik}\frac{\partial a^k}{\partial x^j} \right) dx^i \otimes dx^j,$$

will enable the reader to see his way through the estimates. The notation in (9.16) should now be clear; $\mathcal{D}^*\Phi_{\mu,b}$ is the solution to

$$\mathcal{D}\mathcal{D}^*\Phi_{\mu,b} = -P_-(L_{X_{\mu,b}}F_D)$$

guaranteed by (7.19).

The proof of (9.17) proceeds in two stages. First we show that

$$\langle\mu, b\rangle \mapsto d\mathcal{B}_D(L_{X_{\mu,b}}D)$$

is still close to the identity map.

PROPOSITION 9.21. *The estimate*

$$d\mathcal{B}_D(L_{X_{\mu,b}}D) = \langle\mu, b\rangle + O(\lambda^2)|\langle\mu, b\rangle|$$

holds.

For the rest of this section, all norms refer to the manifold metric g_1.

PROOF: Note first that we can work entirely in $B_{4\lambda}$, since $\mathcal{B}$ is defined from the data there (8.12). Let $F = F_D$, $X = X_{\mu,b}$, $\omega = \omega(D) = -tr(F \wedge *F) = |F|^2 * 1$, and recall that $\mathcal{B}(D) = \mathfrak{b}(\omega, g)$. Thus $d\mathcal{B}_D(L_X D) = d_\omega\mathfrak{b} \circ d\omega_D(L_X D)$. Now

$$d\omega_D(L_X D) = -2tr(F \wedge *L_X F).$$

By comparison,

$$L_X\omega = -2tr(F \wedge *L_X F) - tr(F \wedge L_X(*)F).$$

Hence,

$$(9.22) \qquad d\mathcal{B}_D(L_X D) = d_\omega\mathfrak{b}(\omega, g)(L_X\omega + tr(F \wedge L_X(*)F)).$$

Our construction in the introduction to this section was designed to give

$$(9.23) \qquad\qquad d_\omega\mathfrak{b}(\omega, \delta)(L_{X_{\mu,b}}\omega) = \langle\mu, b\rangle$$

(cf. (9.12)). Also, by (8.28),

$$(9.24) \qquad\qquad \|d_\lambda\mathfrak{b}(\omega, g) - d_\omega\mathfrak{b}(\omega, \delta)\| = O(\lambda^3).$$

The convergence (8.33) in the normalized model II to the standard instanton $\hat{I}_\lambda$ (8.26) for λ sufficiently small implies

$$(9.25) \qquad \|L_X\omega\|_{L^1} \sim \|L_X\hat{I}_\lambda\|_{L^1} = O\left(\frac{1}{\lambda}\right)|\langle\mu,b\rangle|$$

by explicit computation. Combining (9.23)–(9.25),

$$
\begin{aligned}
(9.26) \qquad d_\omega\mathfrak{b}(\omega,g)(L_X\omega) &= [d_\omega\mathfrak{b}(\omega,\delta) + (d_\omega\mathfrak{b}(\omega,g) - d_\omega\mathfrak{b}(\omega,\delta))]\,(L_x\omega) \\
&= \langle\mu,b\rangle + O(\lambda^2)|\langle\mu,b\rangle|.
\end{aligned}
$$

The other term in (9.22) is estimated in balls of radius 4λ by (8.28)

$$
\begin{aligned}
(9.27) \qquad |d_\omega\mathfrak{b}(\omega,g)(tr\,F \wedge L_X(*)F)| &\le \|d_\omega\mathfrak{b}(\omega,g)\|\,\|F \wedge L_X(*)F\|_{L^1}, \\
&= O(\lambda)\|F\|_{L^2}\|L_X(*)\|_{L^\infty}.
\end{aligned}
$$

Here $* = *(g)$, and in exponential coordinates on $B_{4\lambda}$,

$$\| * (g) - *(\delta)\|_{C^0} = O(\lambda^2),$$

$$\| * (g) - *(\delta)\|_{C^1} = O(\lambda).$$

Now X is conformal relative to δ on $B_{4\lambda}$, and $*$ is conformally invariant on 2-forms. Hence $L_X(*(\delta)) = 0$. Now we compute

$$(9.28) \qquad \|L_X(*(g))\|_{L^\infty} = O(\lambda)|\langle b,\mu\rangle|,$$

and the proposition follows from (9.22), (9.26)–(9.28).

The proof of (9.17) is completed by showing that the contribution from Taubes' projection in (9.16) is small. Recall that this term solves the equation

$$\mathcal{D}\mathcal{D}^*\Phi_{\mu,b} = -P_-(L_X F).$$

PROPOSITION 9.29. *We have*

$$d\mathcal{B}_D(\mathcal{D}^*\Phi_{\mu,b}) = O(\lambda)|\langle\mu,b\rangle|.$$

PROOF: By (8.30), for $\Phi = \Phi_{\mu,b}$,

$$\|d\mathcal{B}_D(\mathcal{D}^*\Phi)\| \le c_{28}\lambda\|\mathcal{D}\mathcal{D}^*\Phi\|_{L^2}.$$

But our $\mathcal{H}(D)$ norm (7.17) bounds this,

$$\|D\mathcal{D}^*\Phi\|_{L^2} \le c_{36}\|\Phi\|_{\mathcal{H}(D)},$$

and the estimate in (7.19) is

$$\|\Phi\|_{\mathcal{H}(D)} \le c_{18}\|P_-(L_X F)\|_{L^2}.$$

To prove the proposition, then, it suffices to bound $\|P_-(L_X F)\|_{L^2}$ by $|\langle \mu, b\rangle|$.

Let ψ_t be the one-parameter group generated by X. Then $\psi_t^*(D)$ is self-dual in the metric $\psi_t^*(g)$, which is expressed infinitesimally by

$$P_-(L_X F) = -(L_X P_-)F.$$

Finally, we are left to bound $\|(L_X P_-)F\|_{L^2}$. We remark that the conformal invariance of $*$ on 2-forms implies

$$|L_X P_-| \le c_{37}|\pi(L_X g)|,$$

where π is projection onto the "tracefree" piece.

For translations ($\mu = 0$) there is no problem, since

$$\|L_{X_b} P_-\|_{L^\infty} \le c_{38}|b|,$$

and

$$\|F\|_{L^2} = 8\pi^2.$$

But for dilations there is an extra factor of $1/\lambda$,

$$|L_{X_\mu} P_-|(x) = O(|x|^2)\frac{\mu}{\lambda},$$

and we must work a little. Here is where our decay estimate (9.8),

$$|F|(x) \le \frac{c_{39}\lambda^\gamma}{|x|^{2+\gamma}}, \qquad K\lambda \le |x| \le \rho,$$

enters the argument. Now

$$\|P_-(L_{X_\mu} F)\|_{L^2}^2 \le c_{40}\left\{ \int_0^{K\lambda} r^4 \left(\frac{\mu}{\lambda}\right)^2 |F|^2 * 1 + \int_{K\lambda}^\rho r^4 \left(\frac{\mu}{\lambda}\right)^2 \frac{\lambda^{2\gamma}}{r^{4+2\gamma}} r^3 dr \right\}$$

$$= O(\lambda^2) + O(\lambda^{2\gamma-2})$$

$$= 0(\lambda^{2\gamma-2}).$$

Choosing $\gamma \ge 1$ in (9.8), the result follows.

Finally, these two estimates show that for λ sufficiently small, $\|d\mathcal{B}_D \circ \Sigma_{\mu,b}\| > 0$ for all $\langle \mu, b\rangle$; that is, $d\mathcal{B}_D \circ \Sigma$ is invertible as claimed.

Exponential Gauges.

Now that we know $\overline{B}$ is a covering map (cf. the introduction to this chapter), we turn to the proof that $\overline{B}$ is 1:1. As a preliminary step we use our decay estimate in the entire neck region to produce a gauge in which $D = d + A$ with $|A|$ small. The technical device we need is an *exponential gauge* [U2], and in this section we explain it first in the general context of a vector bundle ξ with connection D and metric over a Riemannian manifold M. The idea is that if $|F| = |F_D|$ is small, then we can locally and geometrically choose a gauge with $|A|$ small. (Previously (8.2) we did this with PDE.) For connections on the tangent bundle, this is accomplished by using geodesic normal coordinates. When we work in extrinsic bundles, we must translate along radial geodesics in the bundle. In fact, the construction amounts to taking geodesic normal coordinates in the total space of the bundle, made in a Riemannian manifold by the fiber metric, base metric, and connection.

Thus fix $p \in M$ and construct a local trivialization of ξ by identifying $\xi_{\exp_p(tX)}$ with ξ_p via parallel translation along the radial geodesic $\{\exp_p(sX), 0 \le s \le t\}$ for each unit $X \in T_p M$ and small t. Fix a frame at p. This introduces the freedom to rotate by a fixed element of the structure group G, just as one has a free constant linear transformation in choosing geodesic framings of the tangent bundle. Then $D = d + A$ with $A_r = 0$ in geodesic polar coordinates $\langle r, \theta \rangle$, $\theta \in S^{n-1}$. Now, in somewhat poetic notation,

$$F_{r\theta} = \frac{\partial A_\theta}{\partial r} - \frac{\partial A_r}{\partial \theta} + [A_r, A_\theta]$$
$$= \frac{\partial A_\theta}{\partial r},$$

so that

$$(9.30) \qquad A_\theta(r_0, \theta_0) = \int_0^{r_0} F_{r\theta}(r, \theta_0)\, dr.$$

We have constructed a gauge in which the connection is expressible in terms of the curvature, and so (9.30) can be used to estimate A in terms of F, if we are careful to include the base metric $dr^2 + r^2(1 + O(r^2))d\theta^2$. Restricting to a region bounded away from the cut locus, these metric contributions are bounded, and

$$(9.31) \qquad \|A\|_{L^\infty(B_r)} \le c_{41}\|F\|_{L^\infty(B_r)},$$

with $c_{41} = c_{41}(r)$ blowing up as we approach the cut locus. When carried out on geodesics normal to a submanifold S on which a gauge is specified, the result is termed a *transverse gauge*, and then (9.31) becomes

$$(9.32) \qquad \|A\|_{L^\infty(S_r)} \leq \|A\|_{L^\infty(S)} + c_{42}(r)\|F\|_{L^\infty(S_r)},$$

where now c_{42} blows up as we approach focal points. Here S_r is a tubular neighborhood of S in M.

Exponential (or transverse) gauges centered at different points can be patched together when the intersection is simple geometrically. We illustrate with the following

PROPOSITION 9.33 [**U2**]. *Let D be a connection on a bundle over S^n. If $\|F\|_{L^\infty(S^n)}$ is sufficiently small, then there exists a global gauge on S^n for which $D = d + A$ with*

$$\|A\|_{L^\infty(S^n)} \leq c_{43}\|F\|_{L^\infty(S^n)},$$

and c_{43} is a constant independent of D.

PROOF: Construct exponential gauges from the south pole and north pole, each extending slightly past the equator. The connection forms A^0, A^∞ thus obtained both satisfy (9.31),

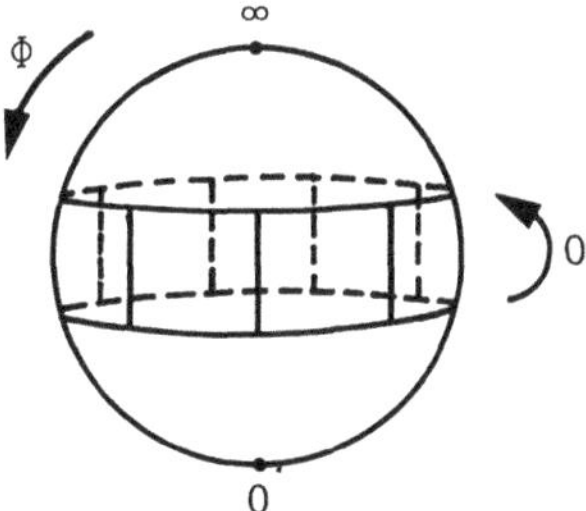

where the constant c_{41} depends only on the geometry of the sphere. On the intersection, $D = d^0 + A^0 = d^\infty + A^\infty$, so that A^0 and A^∞ are related by a gauge change

$$(9.34) \qquad s^{-1}d^0 s = A^0 - A^\infty$$

By construction, exponential gauges vanish in the "radial direction", which in this case is given by the polar angle ϕ, whence

$$\frac{\partial s}{\partial \phi} = s(A_\phi^\infty - A_\phi^0) = 0.$$

Thus $s = s(\theta)$ is a function only of the equatorial variables. Assuming for convenience that s is unitary, we estimate

$$(9.35) \qquad \|ds\|_{L^\infty} \leq 2c_{41}\|F\|_{L^\infty}$$

from (9.31) and (9.34). We now take advantage of the freedom to rotate by a fixed element of the gauge group to arrange that $s(\theta_0) = 1$ for some θ_0. Then integrating (9.34),

$$(9.36) \qquad \|s - 1\|_{L^\infty} \leq c_{41}\pi\|F\|_{L^\infty}.$$

If $\|F\|_{L^\infty}$ is sufficiently small (the yardstick here is determined by the geometry of G), we can write

$$s(\theta) = \exp(u(\theta))$$

for some $u : S^{n-1} \rightarrow \mathfrak{g}$, $\mathfrak{g}$ the Lie algebra of G, and from (9.35) and (9.36),

$$(9.37) \qquad \|u\|_{L_1^\infty} \leq c_{44}\|s - 1\|_{L_1^\infty} \leq c_{41}c_{44}\pi\|F\|_{L^\infty},$$

where $c_{44} = c_{44}(G, \mathrm{dist}(s(\theta), 1))$. Let $\beta(\phi)$ be a cutoff function

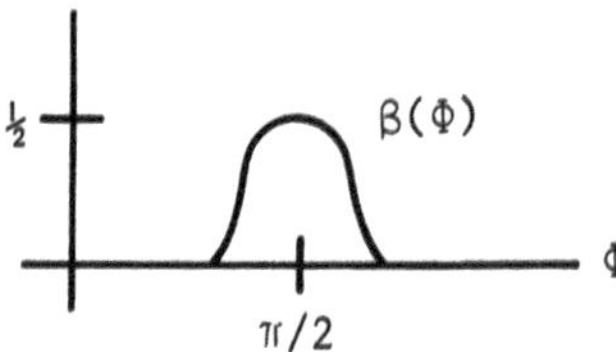

on the overlap such that $\beta\left(\frac{\pi}{2}\right) = \frac{1}{2}$, $|d\beta|$ bounded. Using

$$s^0(\phi, \theta) = \exp(\beta(\phi)u(\theta)),$$
$$s^\infty(\phi, \theta) = \exp(-\beta(\phi)u(\theta)),$$

we can match gauges to write $D = d + A$ globally with

$$A = A^0 - (s^0)^{-1}(d^0 s^0) = A^\infty - (s^\infty)^{-1}(d^\infty s^\infty).$$

The estimate in the proposition follows easily from (9.31) and (9.37).

This technique can also be used to patch transverse gauges on a cylinder, where again the geometry of the intersection is simple.

The general patching argument does not rely on any properties of the intersection.

Returning to our concentrated instanton D, we construct a good gauge on the neck. Introduce $\bar{\tau}_\lambda = (\tau_0 + \tau_\lambda)/2$.

PROPOSITION 9.38. *For $K \geq K_2$ there exists a gauge on N_λ such that $D = d + A$ with*

$$|A|(\tau, \theta) \leq \frac{c_{35}\nu e^{-\bar{\tau}_\lambda}}{\gamma}\left(c_{45} + e^{-\gamma|\bar{\tau}_\lambda - \tau|}\right).$$

In particular,

$$\|A\|_{L^\infty(N_\lambda)} \leq c_{46}\nu.$$

PROOF: Since the metric on N_λ is very close to the cylinder metric, we use the cylinder metric at the expense of a small change in our constants. Apply (9.33) to the S^3 at $\tau = \bar{\tau}_\lambda$ to obtain a gauge in which

$$(9.39) \qquad\qquad |A|(\bar{\tau}_\lambda, \theta) \leq c_{35}c_{43}\nu e^{-\gamma\bar{\tau}_\lambda}$$

by the decay estimate (9.8). Now extend to a transverse gauge ($A_\tau = 0$ on N_λ (there are no focal points to worry about). Then, as in (9.30),

$$A_\theta(\tau, \theta_0) = A_\theta(\bar{\tau}_\lambda, \theta_0) + \int_{\bar{\tau}_\lambda}^{\tau} F_{\tau\theta}(\sigma, \theta_0)d\sigma,$$

from which the proposition follows by integration using (9.8) and (9.39).

COROLLARY 9.40. $\|A\|_{L^4(N_\lambda)} \leq c_{47}\nu.$

Note that since $\nu \to 0$ as $K \to \infty$, for large K the bounds (9.38) and (9.40) become arbitrarily small.

PROOF: Integrate (9.38).

Connectivity of the Collar.

Our description of a concentrated instanton D_λ is now complete. Define the norm

$$(9.41) \qquad \|A\|_\Omega = \|A\|_{L^\infty(\Omega)} + \|A\|_{L^4(\Omega)}$$

on $\Omega^1(\operatorname{ad}\eta)$. Recall that $\|\ \|_{L^4}$ is conformally invariant on 1-forms, but $\|A\|_{L^\infty}$ depends on the metric, and in (9.41) we use the blown-up metric g_3. On I_λ and N_λ this is sufficiently close

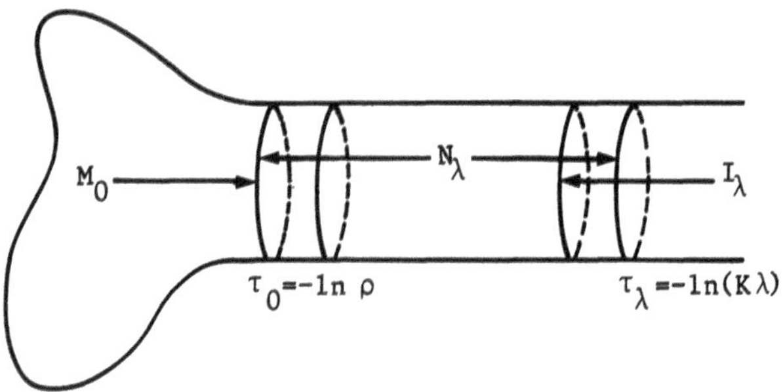

to the cylinder metric that we will ignore the distinction. We collect our results as follows:

$$(9.42) \qquad \begin{aligned} (s_\lambda^I)^* D_\lambda &= D_\lambda^{S^4} + A_\lambda^I && \text{on } I_\lambda, \\ (s_\lambda^N)^* D_\lambda &= d + A_\lambda^N && \text{on } N_\lambda, \\ (s_\lambda^M)^* D_\lambda &= d + A_\lambda^M && \text{on } M_0, \end{aligned}$$

where

$$(9.43) \qquad \lim_{\lambda \to 0} \|A_\lambda^I\|_{I_\lambda} = 0,$$

$$(9.44) \qquad \lim_{\lambda \to 0} \|A_\lambda^N\|_{N_\lambda} \le \frac{c_{30}(c_{46} + c_{47})}{K^2},$$

$$(9.45) \qquad \lim_{\lambda \to 0} \|A_\lambda^M\|_{M_0} = 0.$$

Here $D_\lambda^{S^4}$ is the basic concentrated instanton of scale λ, but transferred to our blown-up model, and d is the trivial connection in some gauge. We have explicitly included the gauge transformations in (9.42) for clarity. Equation (9.43) follows from the convergence to the standard instanton (8.33), where in transferring L^∞ estimates relative to g_2 into L^∞ estimates relative to g_3

on $B_{K\lambda}$, the acquired scale factor is bounded by K. Also, (9.44) follows from our transverse gauge estimates (9.38) and (9.40) in view of (9.4), and (9.45) follows from (8.35).

We are in a position to complete our proof that $\overline{\mathcal{B}}$ is a diffeomorphism. Let $\overline{D}_\lambda$, $\overline{D}'_\lambda$ be curves in $\mathcal{M}$ such that $\overline{\mathcal{B}}(\overline{D}_\lambda) = \overline{\mathcal{B}}(\overline{D}'_\lambda) = \langle \lambda, x \rangle$ for some fixed $x \in M$, with λ varying in an interval.

PROPOSITION 9.46. *For $\lambda \leq \lambda_7$, $K \geq K_3$ there exist lifts D_λ, D'_λ whose difference A_λ satisfies*

$$\lim_{\lambda \to 0} \|A_\lambda\|_{L^4} \leq \frac{c_{48}}{K^2}.$$

The Collar Theorem will then follow from (7.30).

PROOF: By (9.43)–(9.45) and the triangle inequality, there exist lifts D_λ, D'_λ and gauge transformations s^N_λ, s^M_λ on N_λ, M_0 such that

$$
\begin{aligned}
D'_\lambda &= D_\lambda + A^I_\lambda & &\text{on } I_\lambda, \\
(s^N_\lambda)^* D'_\lambda &= D_\lambda + A^N_\lambda & &\text{on } N_\lambda, \\
(s^M_\lambda)^* D'_\lambda &= D_\lambda + A^M_\lambda & &\text{on } M_0,
\end{aligned}
\tag{9.47}
$$

where

$$\lim_{\lambda \to 0} \|A^I_\lambda\|_{I_\lambda} = 0, \tag{9.48}$$

$$\lim_{\lambda \to 0} \|A^N_\lambda\|_{N_\lambda} \leq \frac{2c_{30}(c_{46} + c_{47})}{K^2}, \tag{9.49}$$

$$\lim_{\lambda \to 0} \|A^M_\lambda\|_{M_0} = 0. \tag{9.50}$$

Our job is to patch again, just as in (9.33), only now we are not necessarily dealing with exponential gauges. We illustrate the argument briefly on $I_\lambda \cap N_\lambda$. First, from (9.47) we see that on this intersection

$$D'_\lambda s = s(A^N_\lambda - A^I_\lambda), \tag{9.51}$$

where we denote $s = s^N_\lambda$. If K is chosen sufficiently large, the right hand side is arbitrarily small. Also D'_λ is almost flat on $I_\lambda \cap N_\lambda$, so that (9.51) gives an estimate on ds. Furthermore, rotating s by a constant element of $SU(2)$, we can arrange that $s = 1$ somewhere in $I_\lambda \cap N_\lambda$ without affecting the estimates. Then $\|s - 1\|_{L^\infty_1(I_\lambda \cap N_\lambda)}$ is small, and we can write $s = e^u$ as in (9.33). Define a cutoff function φ,

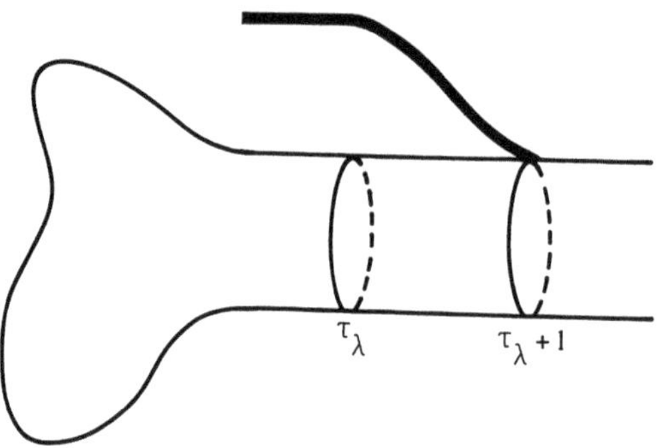

and replace s_λ^N in (9.47) by $e^{\varphi u}$. Now

$$(e^{\varphi u})^* D'_\lambda = D_\lambda + t^{-1} D_\lambda t + t^{-1} A_\lambda^N t \text{ on } N_\lambda$$

for $t = e^{(\varphi - 1)u}$, and a lift for $\overline{D}'_\lambda$ is consistently defined on $I_\lambda \cup N_\lambda$ by

$$\tilde{D}'_\lambda = \begin{cases} D'_\lambda & \text{on } I_\lambda, \\ (e^{\varphi u})^* D'_\lambda & \text{on } N_\lambda. \end{cases}$$

Estimates of the form (9.48) and (9.49) still hold because of our L_1^p control on u, $p = 4, \infty$, and since we can take $|d\varphi|$ bounded. Repeat this patching argument on $N_\lambda \cap M_0$ to obtain a global gauge. The proposition follows by collecting all of the estimates.

We now prove that the fibers of $\overline{B}$ are connected.

THEOREM 9.52. *For $\lambda \leq \lambda_8$ and $K \geq K_4$, we have $\overline{D}_\lambda = \overline{D}'_\lambda$.*

PROOF: Fixing K large to ensure the hypothesis of (7.30), we obtain a path $D_{t,\lambda}$ with $D_{0,\lambda} = D_\lambda$ and $D_{1,\lambda} = D'_\lambda$. Furthermore, convergence to the standard instanton (8.33) on I_λ ensures

$$\lim_{\lambda \to 0} \|F'_\lambda - F_\lambda\|_{L^2(I_\lambda)} = 0,$$

and

$$\lim_{\lambda \to 0} \|F'_\lambda - F_\lambda\|_{L^2(M_0)} = 0$$

by (8.35). Also, our decay estimate (9.8), integrated on N_λ, implies

$$\|F'_\lambda - F_\lambda\|_{L^2(N_\lambda)} \leq \frac{c_{49}}{K^2}.$$

So by the curvature estimate in (7.30),

$$\|F_{t,\lambda} - F_\lambda\|_{L^2(M)} \leq \frac{c_{50}}{K^2}$$

remains small if $\lambda \leq \lambda_8$, say. Since concentrated and unconcentrated curvature differ in L^2 by almost $16\pi^2$,

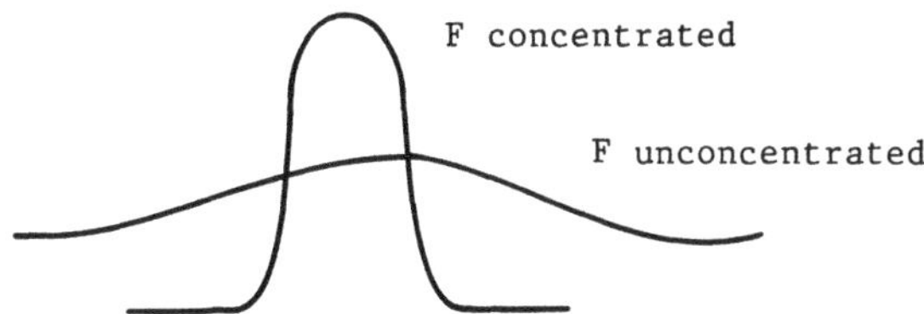

the path $\{\overline{D}_{t,\lambda}, 0 \leq t \leq 1\}$ remains in the collar $\mathcal{M}_{\lambda_8}$ for K large. Also, $\mathrm{dist}(x(\overline{D}_{t,\lambda}), x(\overline{D}_\lambda)) < \lambda$, because $B_\lambda(x(\overline{D}_\lambda))$ contains roughly $4\pi^2$ worth of L^2 energy, and so concentrated instantons whose curvatures are close have

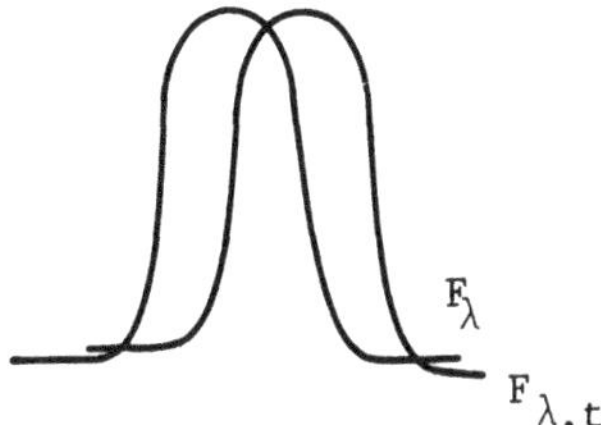

close centers. Therefore, the path $\overline{D}_{t,\lambda}$ in $\mathcal{M}_{\lambda_8}$ is short. Since $\overline{\mathcal{B}}$ is a local diffeomorphism (9.18), we see that $\overline{D}_\lambda = \overline{D}_{\lambda'}$.

Finally, $\overline{\mathcal{B}}$ is a 1:1 local diffeomorphism, hence a global diffeomorphism, and the proof of (9.1), as well as the proof of Donaldson's Theorem, is complete at last.

We remarked at the end of §2 that to study the moduli space of self-dual connections on a G bundle over M we impose three topological conditions on M and G—we require that the intersection form ω be positive definite, the first Betti number b_1 vanish, and the dimension of G be 3—and that there is real trouble if we relax any of these constraints. The differential topologists Ronald Fintushel and Ronald Stern noticed that for $G = SO(3)$, i.e., for oriented real three dimensional vector bundles, a theorem different from Donaldson's can be obtained. Their nonsmoothability result holds for compact oriented 4-manifolds with almost any finite fundamental group, but not all intersection pairings are allowed. (However, their proof does apply to $E_8 \oplus E_8$, and then the existence of fake $\mathbf{R}^4$ follows as before.) One advantage to their approach is that the analysis is much easier. Since their results have important ramifications for 3-manifold topology, we include an "easy" case of their theorem in this chapter. The difficulties in harder cases are not in the analysis, but arise mostly from the number theory of the intersection form, and we provide enough information so that the reader can fill in these details.

The Moduli Space for $SO(3)$ Bundles.

Let us first recall some topological facts about $SO(3)$ bundles (cf. Appendix E). There are two characteristic classes which classify $SO(3)$ bundles ξ over compact oriented 4-manifolds. These are the second Stiefel–Whitney class $w_2(\xi) \in H^2(M; \mathbf{Z}_2)$ and the first Pontrjagin class $p_1(\xi) \in H^4(M; \mathbf{Z})$. We denote the Pontrjagin number $p_1(\xi)[M]$ by ℓ, and have the Chern–Weil formula

$$(10.1) \qquad \ell = p_1(\xi)[M] = \frac{-1}{4\pi^2} \int_M tr(F \wedge F),$$

where F is the curvature of a connection on ξ. Note that $\ell \geq 0$ if F is self-dual. If $w_2(\xi)$ vanishes, then ξ can be realized as the (real) adjoint bundle of an $SU(2)$ bundle η, and in that case $\ell = 4k$, where k is minus the second Chern number of η. (So to get moduli spaces which do not arise from $SU(2)$ bundles, we need $w_2(\xi) \neq 0$.) Note that ad $\xi \cong \xi$ for $SO(3)$ bundles. Suppose now that ξ is a reducible (or split) bundle. This means

$$\xi = \lambda \oplus \varepsilon \qquad \text{(orthogonal direct sum)},$$

where λ is an $SO(2) = U(1)$ bundle, i.e., λ is an oriented real 2-plane bundle = complex line bundle, and ε is a trivial oriented real line bundle. Recalling that w_2, p_1 are *stable* characteristic classes, we compute $[MS]$.

(10.2)
$$w_2(\xi) = w_2(\lambda) = c_1(\lambda)_{\text{mod } 2} \in H^2(M; \mathbb{Z}_2),$$
$$p_1(\xi) = p_1(\lambda) = c_1(\lambda)^2 - 2c_2(\lambda) = c_1(\lambda)^2 \in H^4(M; \mathbb{Z}).$$

We now study the moduli space $\mathcal{M}$ of self-dual connections on ξ. The Atiyah–Singer Index Theorem computes the dimension of $\mathcal{M}$ as

(10.3)
$$\dim \mathcal{M} = 2\ell - 3(1 - b_1 + b_2^-).$$

(This is formula (2.29).) We mimic our discussion of Donaldson's Theorem (§2).

THEOREM 10.4. *Let ξ be an $SO(3)$ bundle over a compact oriented 4-manifold with positive definite intersection form ω and vanishing first Betti number. Assume $\ell = p_1(\xi)[M] = 2$. Then the moduli space $\mathcal{M}$ of self-dual connections on ξ modulo gauge transformations has the following properties.*

I. *Let m be half the number of solutions to*

(10.5)
$$\omega(\alpha, \alpha) = 2,$$
$$\alpha_{\text{mod } 2} = w_2(\xi).$$

   Then for almost all metrics on M, there exist $p_1, \ldots, p_m \in \mathcal{M}$ such that $\mathcal{M} - \{p_1, \ldots, p_m\}$ is a smooth one dimensional manifold.

II. *For almost all metrics on M, there exist neighborhoods $\mathcal{O}_{p_i}$ of p_i so that $\mathcal{O}_{p_1} \cong$ a ray (i.e., a cone on a point $= \mathbb{C}/S^1$).*

III. *$\mathcal{M}$ is compact.*

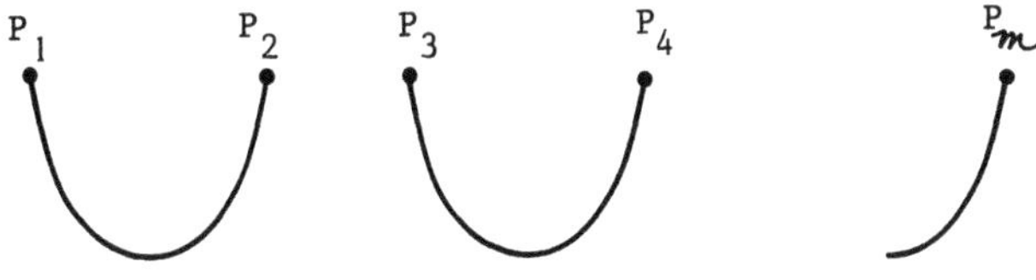

In fact, condition I provides the only complication in that the number of solutions to (10.5) is slightly difficult to count if $H^2(M;\mathbb{Z})$ has torsion. For this reason, we impose the hypothesis $H_1(M;\mathbb{Z}) = 0$ (cf. $(E.1)$), and leave the case $H_1(M;\mathbb{Z}) \neq 0$ for our readers to do as an exercise. The precise condition needed is that $H_1(M;\mathbb{Z})$ has no 2^r-torsion for $r \geq 2$. Here, as in Donaldson's Theorem, the singular points (in this case boundary points) correspond to split bundles. One must be careful to check which split Yang–Mills connections are equivalent under the action of gauge transformations.

We now state the version of the Fintushel–Stern Theorem that we will prove in detail.

THEOREM 10.6 (Fintushel and Stern [**FS**]). *Let M be a compact oriented 4-manifold with positive definite intersection form ω and $H_1(M;\mathbb{Z}) = 0$. If there exists $\alpha \in H^2(M;\mathbb{Z})$ satisfying $\omega(\alpha,\alpha) = 2$ with $\alpha \neq \beta + \gamma$, where $\omega(\beta,\beta) = \omega(\gamma,\gamma) = 1$, then M is not smoothable.*

In particular, $E_8 \oplus E_8$ satisfies the hypothesis, and so is not smoothable. However, there are many positive definite intersection forms which do not take on the value of 2. The 24 dimensional Leech lattice, for example, has minimal vectors of length 4, and the tensor product $E_8 \otimes \cdots \otimes E_8$ (s times) takes on a minimal value of 2^s. So this theorem does not cover all the cases of Donaldson's Theorem. We remark that Fintushel and Stern have a separate argument to cover intersection pairings whose minimal value is 3. (One complication for higher values is that $\mathcal{M}$ is no longer compact.)

We now prove Theorem 10.6 assuming properties I–III of the moduli space.

PROOF OF THEOREM 10.6: Let λ be the $SO(2)$ bundle with $c_1(\lambda) = \alpha$, and $\xi = \lambda \oplus \varepsilon$ the corresponding $SO(3)$ bundle. Then for a generic metric on M the moduli space $\mathcal{M}$ is a compact 1-manifold with m boundary points, where m is half the number of solutions to (10.5). We complete the proof by showing that $\pm\alpha$ are the only solutions to these equations. For then $\mathcal{M}$ has only one boundary point and is compact, a clear impossibility. Now any other solution to (10.5) is of the form $\alpha + 2\beta$ with

$$2 = \omega(\alpha + 2\beta, \alpha + 2\beta)$$

$$= 2 + 4\omega(\alpha,\beta) + 4\omega(\beta,\beta),$$

or

$$\omega(\alpha,\beta) + \omega(\beta,\beta) = 0.$$

But then

$$0 \leq \omega(\alpha + \beta, \alpha + \beta) = 2 - \omega(\beta, \beta),$$

from which $\omega(\beta, \beta) \leq 2$. If $\omega(\beta, \beta) = 0$, then $\beta = 0$ since ω is nondegenerate. If $\omega(\beta, \beta) = 2$, then $\omega(\alpha + \beta, \alpha + \beta) = 0$, and now nondegeneracy implies $\beta = -\alpha$. This yields the solution $\alpha - 2\alpha = -\alpha$ which we already noted above. Finally, if $\omega(\beta, \beta) = 1$, then also $\omega(\alpha + \beta, \alpha + \beta) = 1$, so that $\alpha = -\beta + \gamma$ with $\gamma = \alpha + \beta$ and $\omega(-\beta, -\beta) = \omega(\gamma, \gamma) = 1$. This is forbidden by hypothesis, and so $n = 1$ as claimed.

Reducible Connections.

As in Donaldson's Theorem, special solutions to the $SO(3)$ Yang–Mills equations are obtained from splittings $\xi = \lambda \oplus \varepsilon$. The $SO(2)$ bundle λ is given a Yang–Mills connection (whose curvature is $i = \begin{pmatrix} 0 & 1 \\ -1 & 0 \end{pmatrix}$ times the unique harmonic 2-form f representing $c_1(\lambda)$), and ε is given the trivial, flat connection. Since $SO(2) = U(1)$, our discussion of complex line bundles in §2 applies to λ.

PROPOSITION 10.7. *Let m be half the number of solutions to (10.5). Then there exist exactly m split self-dual connections (up to gauge equivalence) in the $\ell = 2\ SO(3)$ bundle ξ with prescribed $w_2(\xi)$.*

Recall that if Tor $H^2(M; \mathbb{Z}) = 0$, we have already proved that $m = 1$ (cf. (10.6)).

PROOF: The bundle ξ splits topologically into $\xi = \lambda \oplus \varepsilon$ if and only if $c_1(\lambda) = \alpha$ satisfies (10.5). Furthermore, the connections described above are Yang–Mills in ξ. We have only to determine which of the connections become identified via the action of gauge transformations.

We first prove that the connections corresponding to α and $-\alpha$ are gauge equivalent. The $SO(2)$ bundle corresponding to $-\alpha$, which we denote $-\lambda$, has the same underlying 2-plane bundle as λ, but has the opposite orientation; that is, although λ and $-\lambda$ are certainly different as $SO(2)$ bundles, they are canonically equivalent as $O(2)$ bundles, i.e., as *unoriented* real 2-plane bundles, simply by reversing orientation. Furthermore, this equivalence takes Yang–Mills connections to Yang–Mills connections (while the curvature if maps to $-if$). Extend to a map $\lambda \oplus \varepsilon \to (-\lambda) \oplus \varepsilon$ by reversing the orientation on ε. Then the total orientation of ξ is preserved, so that this is an $SO(3)$ bundle automorphism (gauge transformation), and our assertion is proved.

Now suppose α_1, α_2 are solutions to (10.5) with $\alpha_1 \neq \pm\alpha_2$, and let $\xi = \lambda_1 \oplus \varepsilon_1 = \lambda_2 \oplus \varepsilon_2$ be the corresponding splittings. The curvatures of our split Yang–Mills connections take the form $F_j = if_j u_j$ $(j = 1, 2)$, where f_j is the unique harmonic form representing α_j, and under the identification ad $\xi \cong \xi$, u_j is the positively oriented unit vector field in ε_j. If the two connections are gauge equivalent, then certainly their curvatures are also gauge equivalent. But gauge transformations only rotate u_j inside ξ, and fix f_j. Therefore F_1 is not equivalent to F_2 unless $f_1 = \pm f_2$. If there is no torsion in $H^2(M; \mathbf{Z})$, then $\alpha_1 = \pm\alpha_2$ and we are reduced to the previous case. Therefore, we can assume that $\alpha_1 \mp \alpha_2$ is (nonzero) pure torsion. If F_1 and F_2 are gauge equivalent, then there is a gauge transformation mapping u_1 to $\pm u_2$, and hence the orthogonal complement λ_1 maps to $\pm\lambda_2$. This implies $\lambda_1 \cong \pm\lambda_2$, which contradicts the assumption $\alpha_1 \neq \pm\alpha_2$. In this case, then, the split Yang–Mills connections are not gauge equivalent.

For completeness, we state the analogue of Theorem 3.1. Note that $SO(3)$ has no center, so that there is no $\mathbf{Z}_2$ in this case. Recall that we say an $SO(3)$ connection $D \in \mathcal{A}$ is *reducible* (or *split*) if $\xi = \lambda \oplus \varepsilon$ and $D = d_1 \oplus d$.

THEOREM 10.8. *Assume that D is not flat. Then the following are equivalent:*

(a) $\mathcal{G}_D \simeq SO(2)$, *where $\mathcal{G}_D \subseteq \mathcal{G}$ is the stabilizer of D;*
(b) $D : \Omega^0(\mathrm{ad}\,\xi) \to \Omega^1(\mathrm{ad}\,\xi)$ *has a nonzero kernel;*
(c) D *is reducible;*
(d) $\mathcal{G}_D \neq 1$.

There is only a minor change in the argument (b) $\Rightarrow$ (c). We identify ad $\xi \cong \xi$. Now if $Du = 0$ for $u \in \Omega^0(\mathrm{ad}\,\xi) = \Omega^0(\xi)$, then since $d|u|^2 = 2(Du, u) = 0$, we can assume $|u| = 1$. Define the bundle ε as the kernel of u. (In ad ξ this is identified with $\mathbf{R} \cdot u$.) The orthogonal complement $\varepsilon^\perp$ is λ, and the isotropy subgroup $\mathcal{G}_D = SO(2) = \{e^{i\theta u}, 0 \leq \theta \leq 2\pi\}$. The rest of the proof proceeds as in Theorem 3.1.

Analytic Details.

While our transversality theorems (§3,§4) and compactness theorems (§8) are stated for $SU(2)$ bundles, the details go over *verbatim* for $SO(3)$ bundles. (We did not realize that there would be a need for these theorems in the $SO(3)$ case.) Both transversality theorems are designed to show that in the elliptic complex

$$0 \to \Omega^0(\mathrm{ad}\,\xi) \xrightarrow{D} \Omega^1(\mathrm{ad}\,\xi) \xrightarrow{P_-D} \Omega^2_-(\mathrm{ad}\,\xi) \to 0,$$

the second cohomology vanishes for almost all metrics. We have just repeated the proof that nonvanishing zeroth cohomology corresponds to geometric splittings of the bundle. We refer the reader to Theorems 3.17, 4.9, and 4.19. Note that these last two theorems prove the existence of irreducible connections near each reducible connection, essentially using the implicit function theorem.

The Compactness Theorem is a little different, so we outline the proof.

THEOREM 10.9. *Let ξ be an $SO(3)$ bundle with $\ell = p_1(\xi)[M] \leq 3$. Then the moduli space of self-dual connections is compact.*

PROOF: Let D_j be a sequence of self-dual connections. By Theorem 8.8, if $\max_{x \in M} |F_{D_j}|(x) \leq B$, then a subsequence of $\{D_j\}$ is gauge equivalent to a convergent sequence of connections, and the points in the moduli space represented by D_j converge. So assume that $\max_{x \in M} |F_{D_j}|(x) \to \infty$. Then the proof of Theorem 8.31 shows that there exist points $x_j \in M$ and numbers $\lambda_j \to 0$ such that $e^*_{\lambda_j, x_j}(D_j) \to I_K$, where I_K is a nontrivial self-dual $SO(3)$ connection on $\mathbf{R}^4$ restricted to the ball of radius K. Again it follows that I_K extends to an $SO(3)$ connection I on S^4, and since $H^2(S^4; \mathbf{Z}_2) = 0$, I lifts to an $SU(2)$ connection. Now by (10.1),

$$\ell = \frac{1}{4\pi^2} \int_M |F_{D_j}|^2 \geq \frac{1}{4\pi^2} \int_{B_K} |F_{I_K}|^2.$$

But for large K,

$$\frac{1}{4\pi^2} \int_{B_K} |F_{I_K}|^2 \approx \frac{1}{4\pi^2} \int_{\mathbf{R}^4} |F_I|^2 = 4k,$$

where k is the topological change of I. Since $\ell \leq 3$, we deduce $k = 0$, whence $F_I \equiv 0$, and the blow-up cannot occur. So the moduli space is compact, as claimed.

APPENDIX A
THE GROUP OF SOBOLEV GAUGE TRANSFORMATIONS

In §3 and §5 we made several technical assertions about the action of $\mathcal{G}_\ell$, the group of Sobolev gauge transformations. We introduced the machinery to prove these statements in §6 when we listed the Sobolev Embedding Theorems, the multiplication properties of Sobolev spaces, and the Composition Lemma. Here we complete the proofs. Recall that if ξ is a vector bundle over a compact domain M, $L_k^p(\xi)$ is the Banach space of sections whose derivatives of order less than or equal to k are in L^p. For $p = 2$, $H_\ell(\xi) = L_\ell^2(\xi)$ is a Hilbert space. The spaces are separable and $C^\infty(\xi)$ is dense in them for $1 \le p < \infty$. They are reflexive if $1 < p < \infty$. Properties of these spaces are listed in §6 as (6.30)–(6.34). We review our basic definitions. A special definition is needed for the group of gauge transformations, since gauge transformations are sections of a fiber bundle.

$$\mathcal{G}_\ell = \{s \in H_\ell(\operatorname{End}\eta) : s^* s = 1 \text{ a.e.}\} \subseteq H_\ell(\operatorname{End}\eta)$$

$$(A.1) \qquad \Omega^i(\xi)_k = H_k(\Lambda^i T^* M \otimes \xi)$$

$$\mathcal{A}_{\ell-1} = \{D_0 + A : D_0 \in \mathcal{A} \text{ and } A \in \Omega^1(\operatorname{ad}\eta)_{\ell-1}\}$$

We fix our manifold M to have dimension 4.

PROPOSTION A.2. $\mathcal{G}_\ell = H_\ell(\operatorname{Aut}\eta)$ *is a Hilbert Lie group with Lie algebra* $H_\ell(\operatorname{ad}\eta)$ *for* $\ell > 2$.

PROOF: We describe first the exponential map, which shows the existence of a manifold structure. At $s \in \mathcal{G}_\ell$, define the exponential for $u \in H_\ell(\operatorname{ad}\eta)$ by

$$(\operatorname{Exp}_s u)(x) = \exp_{s(x)} u(x).$$

This makes sense, since both s and u are continuous and $(\operatorname{Exp}_s u)(x) = \bar{s}(x)$ certainly satisfies $\bar{s}^*(x)\bar{s}(x) = 1$ everywhere, since $\bar{s}(x)$ is continuous in x. Smoothness follows from the Composition Lemma applied to the smooth map exp and the H_ℓ functions $\langle s, u \rangle$. To show that we have a smooth manifold we must prove that the composition $\operatorname{Exp}_t^{-1} \circ \operatorname{Exp}_s$ is smooth and invertible on an open disk in $H_\ell(\operatorname{ad}\eta)$ for t close to s. However,

$$(\operatorname{Exp}_t^{-1} \circ \operatorname{Exp}_s)(u)(x) = \exp_{t(x)}^{-1}(\exp_{s(x)} u(x)),$$

and by the Composition Lemma this map is smooth for $\langle t, u \rangle$ varying over an open set in $\mathcal{G}_\ell \times H_\ell(\mathrm{ad}\,\eta)$. Let σ be $1/3$ the injectivity radius of $SU(2)$; then the inverse $\mathrm{Exp}_s^{-1} \circ \mathrm{Exp}_t$ of $\mathrm{Exp}_t^{-1} \circ \mathrm{Exp}_s$ is well-defined on

$$\{\langle t, u \rangle \in \mathcal{G}_\ell \times H_\ell(\mathrm{ad}\,\eta) : \mathrm{dist}(s(x), t(x)) \leq \sigma, |u(x)| < \sigma\}.$$

Since the H_ℓ topology dominates the C^0 topology, this gives smooth overlaps of charts for t close to s in $\mathcal{G}_\ell$. Here we depend on the Sobolev embedding $H_\ell \subset C^0$ for $\ell > 2$ over a manifold of dimension 4.

The smoothness of the group action is proved the same way. Note in this range H_ℓ is an algebra, so $(s \cdot u)(x) = s(x)u(x)$ defines a bilinear map

$$H_\ell(\mathrm{End}\,\eta) \times H_\ell(\mathrm{End}\,\eta) \to H_\ell(\mathrm{End}\,\eta)$$

which restricts to $H_\ell(\mathrm{Aut}\,\eta) = \mathcal{G}_\ell$. Likewise inversion is the restriction to $\mathcal{G}_\ell$ of the linear map $s \to s^*$ in $H_\ell(\mathrm{End}\,\eta)$.

The group $\mathcal{G}_2$ satisfies some of these properties. However, due to equality in the Sobolev Theorem, $\mathcal{G}_2$ is not a manifold. So smoothness has no meaning and great care must be used in dealing with this group.

PROPOSITION A.3. *$\mathcal{G}_\ell$ acts smoothly on $\mathcal{A}_{\ell-1}$ for $\ell > 2$.*

PROOF: Write $\mathcal{A}_{\ell-1} = \{D_0 + A : A \in \Omega^1(\mathrm{ad}\,\eta)_{\ell-1}\}$ for some fixed $D_0 \in \mathcal{A}$. Now for $D = D_0 + A$,

$$s^*(D) = D_0 + s^{-1}D_0 s + s^{-1}As$$

so that the action of $\mathcal{G}_\ell$ can be written

$$(s, A) \mapsto s^{-1}D_0 s + s^{-1}As.$$

The map $s \mapsto s^{-1}$ is smooth in $\mathcal{G}_\ell$, and the map $s \mapsto D_0 s$ is smooth from $\mathcal{G}_\ell \to \Omega^1(\mathrm{End}\,\eta)_{\ell-1}$. Moreover, in this range $H_{\ell-1}$ is an H_ℓ-module.

PROPOSITION A.4. *The curvature operator*

$$F : \mathcal{A}_{\ell-1} \to \Omega^1(\mathrm{ad}\,\eta)_{\ell-2}$$

is smooth for $\ell > 2$.

PROOF: Locally, using $D = D_0 + A$, $D_0 \in \mathcal{A}$ and $A \in \Omega^1(\mathrm{ad}\,\eta)_{\ell-1}$ as above, we have

$$F(D_0 + A) = F(D_0) + D_0 A + A \wedge A.$$

Clearly the map $A \mapsto D_0 A$ maps $\mathcal{A}_{\ell-1} \to \Omega^1(\mathrm{ad}\,\eta)_{\ell-2}$ linearly, and $F(D_0)$ is smooth. The quadratic term $A \mapsto A \wedge A$ is handled by the Multiplication Lemmas. If $\ell > 3$, $H_{\ell-1}$ is an algebra, and we get $A \mapsto A \wedge A$ maps $H_{\ell-1} \to H_{\ell-1} \subset H_{\ell-2}$. For $\ell = 3$, $H_2 \subset L_1^3$ and $L_1^3 \otimes L_1^3 \subset L_1^2$ in dimension 4 from (6.32) and (6.34).

We also provide the missing part of the proof that the orbit space $\hat{\mathcal{X}}_{\ell-1}$ is Hausdorff.

PROPOSITION A.5. *If $D_n' = s_n^*(D_n)$ for D_n' and D_n in $\mathcal{A}_{\ell-1}$, $\ell > 2$, and $s_n \in \mathcal{G}_1$, then $s_n \in \mathcal{G}_\ell$. Moreover, if D_n' and D_n converge in $\mathcal{A}_{\ell-1}$, then a subsequence $s_{n'}$ converges in $\mathcal{G}_\ell$.*

PROOF: The fact that $s_n \in \mathcal{G}_\ell$ will follow if we proceed to estimate $\|s_n\|_\ell$ in $H_\ell(\mathrm{End}\,\eta)$. We write $D_n' = D + A_n'$ and $D_n = D + A_n$. Then

$$A_n' = s_n^{-1} D s_n + s_n^{-1} A_n s_n,$$

or

$$(A.6) \qquad\qquad D s_n = A_n s_n - s_n A_n'.$$

Note $s_n \in H_1(\mathrm{End}\,\eta) \cap L^\infty(\mathrm{End}\,\eta)$, and so this equation makes sense. Observe that since s_n is unitary almost everywhere and the norm is the trace norm, we have

$$(A.7) \qquad \begin{aligned} d|s_n| &\leq |D s_n| \leq |A_n s_n| + |s_n A_n'| \\ &= |A_n| + |A_n'|. \end{aligned}$$

By the Sobolev Theorem A_n' and A_n are bounded in L^p for $\frac{\ell-1}{4} - \frac{1}{2} + \frac{1}{p} > 0$. Take p any number between 4 and infinity. Now $D s_n$ is bounded in L^p and s_n is bounded in L_1^p.

Go back to (A.6). Use the Sobolev theorem to obtain bounds on A_n' and A_n in L_1^4, which is correct since $\frac{\ell-2}{4} - \frac{1}{2} + \frac{1}{4} \geq 0$ if $\ell \geq 3$. Repeat the Multiplication Theorems, this time observing that L_1^4 is an L_1^p-algebra since $p > 4 = \dim M$. This bounds $D s_n$ in L_1^4, so that $s_n \in L_2^4$. Return to (A.6) and apply the fact that $L_2^2 \subset L_{\ell-1}^2$ is an L_2^4-algebra as $2 \cdot 4 > \dim M = 4$. This estimates $D s_n$ in L_2^2 and s_n in L_3^2.

If $\ell = 3$, we are done. Otherwise we use an induction. Assume s_n is bounded in L_k^2 for $k < \ell$. Then A_n and A_n' are also bounded in $L_k^2 \supset L_{\ell-1}^2$.

Since L^2_k is an algebra for $k > 2$, Ds_n is bounded in L^2_k and s_n is bounded in L^2_{k+1}. Proceed until $k = \ell$.

Now choose a weakly convergent subsequence s'_n in $L^2_\ell(\operatorname{End}\eta)$. By strict inequality in the Sobolev theorems, s'_n converges in $L^q_{\ell-1}$, for $2 < q < 4$. But $L^q_{\ell-1} \otimes L^2_{\ell-1} \to L^2_{\ell-1}$ is continuous since $q(\ell - 1) > 2q > 4$. Consequently $Ds_{n'}$ converges in $L^2_{\ell-1}$, whence $s_{n'}$ converges in L^2_ℓ. Since $\mathcal{G}$ is closed in $H_\ell(\operatorname{End}\eta)$, the limit $s \in \mathcal{G}_\ell$.

Finally, we recall that the group $\mathcal{G}$ is to act on $\Omega^i(\operatorname{ad}\eta)$, but in a Hilbert space setting. Here the action is given by

$$(A.8) \qquad\qquad (s, \Phi) \to s^*\Phi = \operatorname{ad}(s^{-1})\Phi = s^{-1}\Phi s.$$

PROPOSITION A.9. *$\mathcal{G}_\ell$ acts smoothly on $\Omega^i(\operatorname{ad}\eta)_k$ for $-\ell \le k \le \ell$, $\ell > 2$.*

PROOF: We need to show that the map

$$\mathcal{G}_\ell \times \Omega^i(\operatorname{ad}\eta)_k \to \Omega^i(\operatorname{ad}\eta)_k$$

given by $(A.8)$ is smooth. For $k \ge 0$, this follows because H_k is an H_ℓ-algebra, $k < \ell$, $2 < \ell$. It follows for the dual spaces $\Omega^i(\operatorname{ad}\eta)_{-k} \simeq (\Omega^i(\operatorname{ad}\eta)_k)^*$ since the action of s is linear in Φ and self-adjoint.

COROLLARY A.10. *$\mathcal{G}_\ell$ acts eqivariantly with respect to the map*

$$L : \mathcal{A}_{\ell-1} \times \Omega^1(\operatorname{ad}\eta)_{\ell-1} \to \Omega^0(\operatorname{ad}\eta)_{\ell-2} \oplus \Omega^2_-(\operatorname{ad}\eta)_{\ell-2}$$

given by $L(D, A) = D^*A \oplus P_- DA$. Moreover $L(D, \cdot)$ is Fredholm for $D \in \mathcal{A}_{\ell-1}$.

Here we give an alternative, geometric proof of (5.13):

THEOREM B.1. *If M is a compact, oriented, 4-manifold with $H_1(M) = 0$, then*

$$[M, S^3] = \begin{cases} \mathbf{Z}_2 & \text{if } \omega \text{ is even,} \\ 0 & \text{if } \omega \text{ is odd.} \end{cases}$$

As usual, we denote the intersection form of M by ω. Our proof is based on a geometric description of $[M, S^r]$ in any dimension.

Temporarily, let M denote any compact manifold. Then we say that two compact submanifolds N, N' are *cobordant within* M if there exists a compact $Z \subseteq M \times [0, 1]$ such that

$$\partial Z = N \times \{0\} \cup N' \times \{1\}.$$

Cobordism is an equivalence relation on the set of submanifolds of M. We think of the $[0, 1]$ factor as time, and as time evolves the cobordism Z transforms N into N'. There are many refinements of the basic notion of cobordism; relevant here is *framed* cobordism. A *framing* of a submanifold N is a smooth basis of sections $\mathbf{o}_N$ of the normal bundle $\nu_{N \hookrightarrow M}$ to N. Then two framed submanifolds N, N' are *framed cobordant* if there exists a cobordism Z together with a framing $\mathbf{o}_Z$ of $\nu_{Z \hookrightarrow M \times [0,1]}$ which restricts to $\mathbf{o}_N$, $\mathbf{o}_{N'}$ at $M \times \{0\}$, $M \times \{1\}$ respectively.

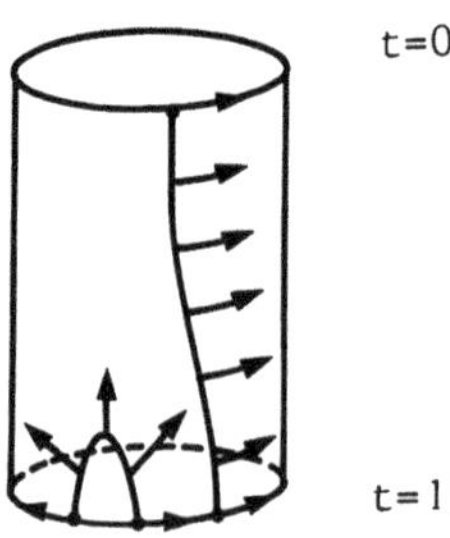

Framed cobordism is again an equivalence relation, and now there is a simple homotopy theoretic description.

THEOREM B.2 (Pontrjagin–Thom). *The equivalence classes of framed sub-manifolds of codimension r are in 1–1 correspondence with $[M, S^r]$.*

The proof can be found in [**M3**]; we are content to construct the correspondence. Suppose $f : M \to S^r$ has $y \in S^r$ as a regular value. Then $N = f^{-1}(y)$ is a smooth submanifold of M, and for each $x \in N$, df_x maps the normal space $T_x M / T_x N$ isomorphically onto $T_y S^r$. So fixing a frame $\mathfrak{o}_y$ at $y \in S^r$, we obtain a framing of N by pullback: $\mathfrak{o}_N(x) = (df_x)^{-1}(\mathfrak{o}_y)$. Conversely, the framed normal bundle to a framed submanifold N can be realized by a trivial tubular neighborhood $N \times \mathbf{R}^r \subseteq M$ for which the

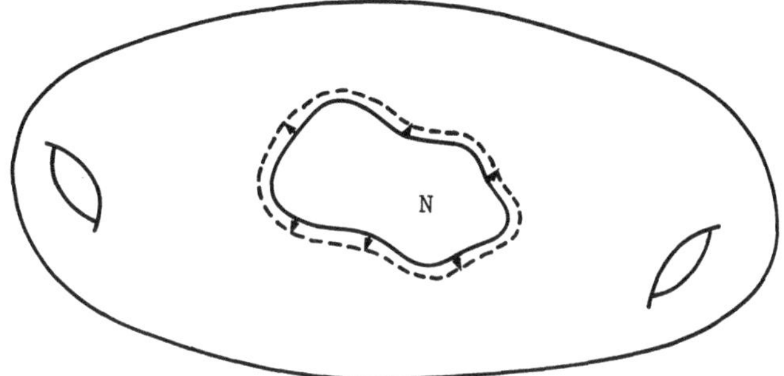

standard frame of $\mathbf{R}^r$ induces $\mathfrak{o}_N$. Modifying the projection $\pi : N \times \mathbf{R}^r \to \mathbf{R}^r$ by a cutoff function β,

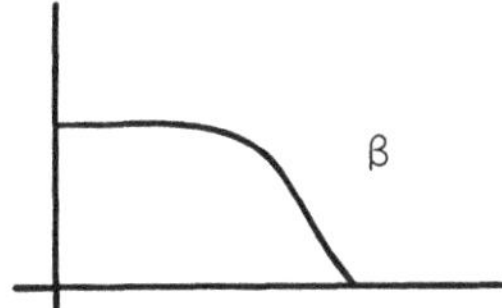

we obtain a map $f : M \to S^r$ defined by

$$f(x) = \begin{cases} \dfrac{\pi(x)}{\beta(|x|)} & \text{if } x \in N \times \mathbf{R}^r \\ \infty & \text{if } x \notin N \times \mathbf{R}^r, \end{cases}$$

where S^r is identified with $\mathbf{R}^r \cup \{\infty\}$ by stereographic projection. The homotopy class $[f] \in [M, S^r]$ is determined uniquely by $\langle N, \mathfrak{o}_N \rangle$.

As a warmup consider the case $\dim M = r$. Then a codimension r submanifold is merely a finite collection of points, and a framing $\mathfrak{o}_x$ at a point $x \in M$ is just a choice of a basis for $T_x M$. If M is oriented, i.e., the first Stiefel–Whitney class $w_1(M)$ vanishes, then we assign

$+1$ to $\mathbf{o}_x$ if it is positively oriented, and -1 if it is negatively oriented. But $N = \langle x, +1 \rangle \cup \langle x', -1 \rangle$ is framed cobordant to $\emptyset$, and so by summing the orientations

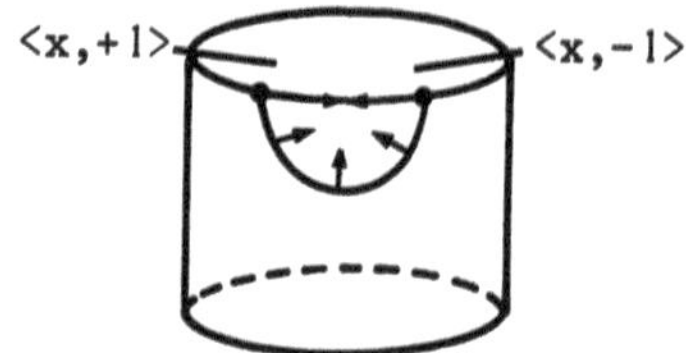

it follows easily from $(B.2)$ that $[M, S^r] \simeq \mathbf{Z}$, the isomorphism being degree. Suppose now that M is nonorientable, and let $S^1 \simeq M$ be a circle for which $\Lambda^r T^* M|_{S^1}$ is a Mobius band. Then by going around the Mobius band, we construct a framed cobordism between $+1$ and -1. In this case $[M, S^r] \simeq \mathbf{Z}_2$, the isomorphism being unoriented degree.

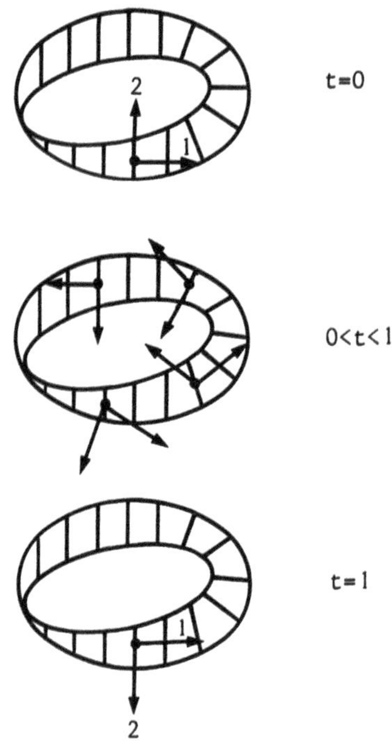

We now compute $[M, S^3]$ for M a compact oriented 4-manifold with $H_1(M) = 0$. By $(B.2)$ this is the set of framed 1-manifolds up to framed cobordism. If we ignore the framing, any compact 1-manifold (= union of circles) is cobordant to $\emptyset$ since $H_1(M) = 0$. The quotient of two framings $\mathfrak{o}_{S^1}$ and $\mathfrak{o}'_{S^1}$ determines a map $S^1 \to SO(3)$ since the normal bundle $\nu_{S^1 \hookrightarrow M}$ is trivial. Let N, N' be two framed circles in M. Then a framed cobordism Z between N and N' defines a homotopy $\mathfrak{o}_N \sim \mathfrak{o}_{N'}$ (the difference is zero), and conversely a homotopy $\mathfrak{o}_N \sim \mathfrak{o}_{N'}$ determines a framed cobordism. We elaborate: if Z is an *unframed* cobordism between N and N', then it is easy to see that $\nu_{Z \hookrightarrow M}$ is trivial, so we can extend the framing of N' over Z. The extended framing restricts to a new framing of N, and we have shown that any framed circle is framed cobordant to a framing of N. Framed cobordisms between different framings of N are exactly maps $N \times [0, 1] \to SO(3)$, i.e., homotopies of loops in $SO(3)$. Hence $[M, S^3]$ is at most $\pi_1(SO(3)) = \mathbb{Z}_2$. If the intersection form ω of M is odd, then there is an oriented surface $\Sigma \subseteq M$ whose reduced Euler class $\chi(\Sigma)_{\text{mod } 2} \in H^2(M; \pi_1(SO(3))) = H^2(M; \mathbb{Z}_2)$ is nonzero (cf. §1). By analogy with the Mobius band cobordism above, construct a framed cobordism between framings $\mathfrak{o}, \mathfrak{o}'$ whose difference is nonzero by having Σ swallow $\mathfrak{o}$ and excrete $\mathfrak{o}'$.

In this case, then, $[M, S^3] = 0$. When ω is even, no such cobordism is possible. In both cases framed 1-manifolds consisting of many framed circles are framed cobordant to a single framed circle by the framed "pair of pants" cobordism.

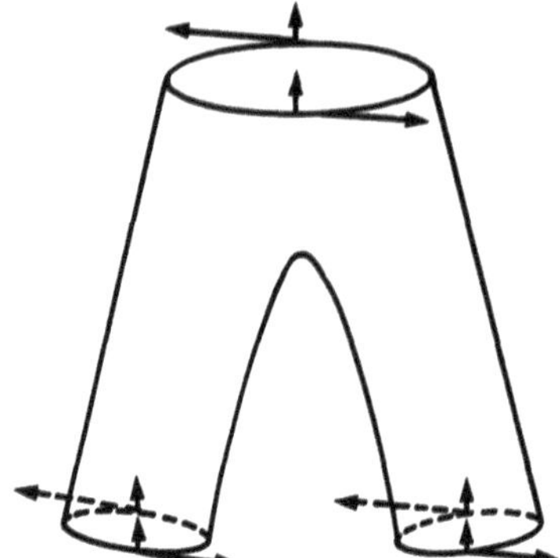

This completes the proof of $(B.1)$.

APPENDIX C
WEITZENBÖCK FORMULAS

The Weitzenböck formulas can be derived intrinsically by working directly on the bundle of frames. This geometric framework emphasizes the $SO(4)$ symmetry, and apart from numerical coefficients, (6.24)–(6.26) follow directly from representation theory. For convenience of exposition we avoid introducing the Dirac operator, although Weitzenböck decompositions correspond to Clifford modules [**ABS**], and hence to Dirac operators. The derivation from that point of view is presented in [**AHS**], [**Pa**], and an alternate exposition is given in [**Bou**].

We begin with some generalities about differential operators on Riemannian manifolds. Let $SO(M)$ denote the principal $SO(n)$ bundle of frames of an oriented n-dimensional Riemannian manifold M. On $SO(M)$ there is a canonical $\mathfrak{so}(n)^*$-valued vertical vector field (i.e., a vertical vector field for each element of $\mathfrak{so}(n)$) encoding the principal bundle structure, and a canonical $\mathbf{R}^n$-valued 1-form which gives the coordinates of tangent vectors. The Levi–Civita connection is an $\mathfrak{so}(n)$-valued 1-form Θ defining projection onto the vertical tangent space. There is a dual description better suited for describing differential operators. Namely, the Levi–Civita connection defines a horizontal $\mathbf{R}^n$-valued vector field ∂. We write $\partial = e^k \otimes \partial_k$, where $\{e^k\}$ is the standard basis of $\mathbf{R}^{n^*}$; then $\partial_k(f)$ is the horizontal lift of the kth element of the frame $f \in SO(M)$. The dual transformation laws

$$(C.1) \qquad \begin{aligned} (R_g)_*(\partial) &= \ g \cdot \partial, \\ (R_g)^*(\Theta) &= \ \mathrm{Ad}\ g^{-1} \cdot \Theta \end{aligned}$$

describe how ∂, Θ change under right translation in a fiber by $g \in SO(n)$. These formulas express the $SO(n)$ symmetry. The Riemannian curvature $\mathcal{R}$, represented on $SO(M)$ by an equivariant map $\mathcal{R} : SO(M) \to \Lambda^2\mathbf{R}^{n^*} \otimes \mathfrak{so}(n)$, is defined by

$$(C.2) \qquad \mathcal{R} = -\Theta([\partial, \partial]),$$

where

$$[\partial, \partial] = e^k \wedge e^\ell \otimes [\partial_k, \partial_\ell].$$

This formalism carries over to extrinsic G bundles P with connection if we work on $SO(M) \times P$. In particular, an (orthogonal) representation ρ :

169

$SO(n) \times G \to \mathrm{Aut}(V)$ defines an associated vector bundle $\tilde{V} = (SO(M) \times P) \times_G V$ with connection ∂ whose curvature is

$$(C.3) \qquad\qquad -\dot{\rho}\Theta([\partial, \partial]).$$

Here $\dot{\rho} : \mathfrak{so}(n) \oplus \mathfrak{g} \to \mathrm{End}(V)$ is the Lie algebra representation induced by ρ. If ρ is a product representation, then $(C.3)$ splits (by a slight abuse of notation) into the sum $\mathcal{R} + F$ of intrinsic and extrinsic curvatures.

An $SO(n)$ equivariant map $A : SO(M) \to \mathrm{Hom}(\mathbf{R}^{n^*}, \mathrm{Hom}(V_1, V_2))$ determines a first order differential operator $A(\partial) = A(e^k)\partial_k$. This acts on a section $\psi : SO(M) \to V_1$ of $\tilde{V}_1$ to give $A(\partial)\psi = A(e^k)(\partial_k\psi) : SO(M) \to V_2$, a section of $\tilde{V}_2$. The map A is the *symbol* of $A(\partial)$, and the adjoint $A(\partial^*)$ has symbol $A^*(e) = -A(e)^*$, the minus sign arising from integration by parts. A sequence of first order symbols

$$(C.4) \qquad\qquad V_0 \xrightarrow{A_1(e)} V_1 \xrightarrow{A_2(e)} V_2 \longrightarrow \cdots \xrightarrow{A_r(e)} V_r$$

determines an *elliptic complex* if $(C.4)$ is exact for each nonzero $e \in \mathbf{R}^{n^*}$. For convenience, collect $V = \oplus V_i$ and $A = \oplus A_i$, and then define

$$(C.5) \qquad\quad L(e, e') = -(A(e)A(e')^* + A(e)^*A(e')).$$

The *Laplacian* $L(\partial^2) = L(e^k, e^\ell)\partial_k\partial_\ell$ associated to $(C.4)$ has as its symbol the symmetric part of L. The simplest Laplacian $\nabla^*\nabla = -\sum_k \partial_k\partial_k$ is associated to the two-step complex

$$(C.6) \qquad\qquad V \xrightarrow{\otimes} V \otimes \mathbf{R}^{n^*},$$

and is defined for any V (cf. (6.23)). There is a *Weitzenböck formula* relating $(C.4)$ and $(C.6)$ precisely when

$$L(e, e) = -\|e\|^2,$$

i.e., when the symbols of $(C.4)$ and $(C.6)$ agree. (Then the map $e \mapsto A(e) - A(e)^*$ induces a Clifford module structure on V.) Let ρ be the representation defining V. The general Weitzenböck formula is

$$
\begin{aligned}
L(\partial^2) &= L(e^k, e^\ell)\partial_k\partial_\ell \\
&= -\sum_k \partial_k\partial_k + \sum_{k<\ell} L(e^k, e^\ell)[\partial_k, \partial_\ell] \\
&= \nabla^*\nabla - \sum_{k<\ell} L(e^k, e^\ell) \circ \dot{\rho}\Theta([\partial_k, \partial_\ell]).
\end{aligned}
$$

$(C.7)$

The correction term is a zeroth order self-adjoint operator involving only the curvature. Transformation laws $(C.1)$ and the $SO(n)$ equivariance of A imply that the correction term in $(C.7)$ is also $SO(n)$ equivariant. This enables us to deduce its form from group theory.

The linear algebra version of our basic complex (6.22) is the symbol sequence

$(C.8)$
$$0 \to \Lambda^0 \mathbf{R}^{4^*} \otimes \operatorname{ad} W \xrightarrow{\varepsilon \otimes \mathrm{id}} \Lambda^1 \mathbf{R}^{4^*} \otimes \operatorname{ad} W$$
$$\xrightarrow{\sqrt{2}P_- \, \varepsilon \otimes \mathrm{id}} \Lambda^2_- \mathbf{R}^{4^*} \otimes \operatorname{ad} W \to 0,$$

where ε is exterior multiplication. Denoting interior multiplication by ι, the associated Laplacians have as symbols the symmetric parts of

$(C.9)$
$$L_0(e^k, e^\ell) = -\iota(e^k)\varepsilon(e^\ell) \otimes \mathrm{id}$$
$$L_1(e^k, e^\ell) = -(\varepsilon(e^k)\iota(e^\ell) + 2\iota(e^k)P_-\varepsilon(e^\ell)) \otimes \mathrm{id}$$
$$L_2(e^k, e^\ell) = -2P_-\varepsilon(e^k)\iota(e^\ell) \otimes \mathrm{id},$$

and the curvature term in $(C.7)$ decomposes into the sum

$(C.10)$
$$\sum_{k<\ell} L(e^k, e^\ell) \circ \mathcal{R}(e_k, e_\ell) \otimes \mathrm{id} + \sum_{k<\ell} L(e^k, e^\ell) \otimes F(e_k, e_\ell),$$

where $\{e_k\}$ is the basis of $\mathbf{R}^4$ dual to $\{e^k\}$. Now $\mathcal{R}(e_k, e_\ell) = L_0(e^k, e^\ell) = 0$ on $\mathbf{R}$ for $k < \ell$, so that $(C.10)$ vanishes for L_0. This proves (6.24). A short computation shows that $L_1(\alpha)$ is nonzero only for self-dual α, and $L_2(\beta)$ is nonzero only for anti-self-dual β. Dropping W momentarily, we then have $SO(4)$ equivariant linear maps

$$L_1 : \Lambda^2_+ \to \Lambda^1 \otimes \Lambda^1,$$
$$L_2 : \Lambda^2_- \to \Lambda^2_- \otimes \Lambda^2_-.$$

In the irreducible decomposition of each of the representations $\Lambda^2_+ \otimes \Lambda^1 \otimes \Lambda^1$ and $\Lambda^2_- \otimes \Lambda^2_- \otimes \Lambda^2_-$, computed using the weight diagram for $\mathfrak{so}(4) = \mathfrak{so}(3) \oplus \mathfrak{so}(3)$,

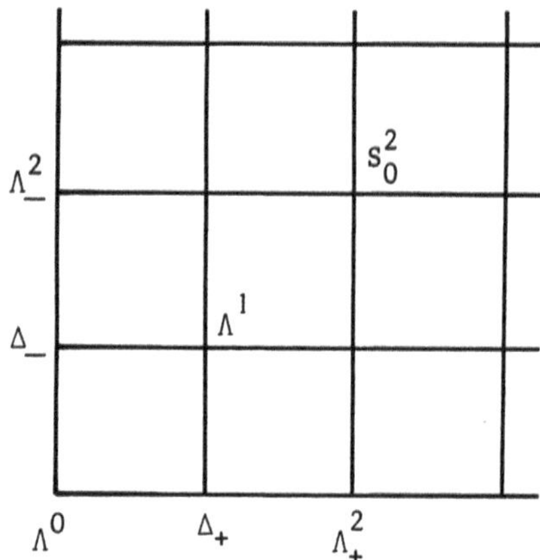

there is only a one dimensional trivial piece. Therefore, up to a scale factor, L_1 and L_2 must be (6.27) and (6.28) respectively. Furthermore, the representation defining ad W is Lie bracket, and we have explained (up to a factor) the extrinsic curvature terms in (6.25) and (6.26). To calculate the intrinsic curvature terms, we examine the first term in $(C.10)$. Note first that the curvature enters linearly, and since $\sum_{k<\ell} L_1(e^k, e^\ell) \circ \mathcal{R}(e_k, e_\ell)$ is an equivariant linear map $\Lambda^1 \to \Lambda^1$, from the decomposition $\mathcal{R} = W^+ \oplus W^- \oplus \mathrm{Ric}_o \oplus R$ of the Riemannian curvature, we see that this map is a linear combination of Ric_o and R. Similarly, $\sum_{k<\ell} L_2(e^k, e^\ell) \circ \mathcal{R}(e_k, e_\ell)$ is an equivariant map $\Lambda^2_- \to \Lambda^2_-$ which must be a linear combination of W^- and R. This analysis gives the formulas (6.25) and (6.26) up to numerical coefficients.

To illustrate the computations involved, we now calculate the intrinsic curvature terms in (6.26), leaving the easier computation of external curvature and of (6.25) to the reader. One could proceed directly from $(C.9)$, but instead we will rederive $(C.7)$ by computing in moving frames. (Alternatively, calculate specific examples to determine the coefficients.) The computation in geodesic normal coordinates is exactly the same. We derive the full Weitzenböck decomposition on 2-forms, and then deduce (6.26). Let θ^1, θ^2, θ^3, θ^4 be a moving frame for which the connection forms vanish at a point, and let X_1, X_2, X_3, X_4 be the dual framing of the tangent space. Then at that point

$$d = \varepsilon(\theta^k)\nabla_{X_k}$$

$$d^* = -\iota(\theta^\ell)\nabla_{X_\ell},$$

and so

$$
\begin{aligned}
\nabla &= dd^* + d^*d \\
&= -(\varepsilon(\theta^k)\iota(\theta^\ell) + \iota(\theta^k)\varepsilon(\theta^\ell))\nabla_{X_k}\nabla_{X_\ell} \\
&= -\sum_k \nabla_{X_k}\nabla_{X_k} - \sum_{k<\ell}(\varepsilon(\theta^k)\iota(\theta^\ell) + \iota(\theta^k)\varepsilon(\theta^\ell))[\nabla_{X_k},\nabla_{X_\ell}] \\
&= \nabla^*\nabla + \sum_{k<\ell}(\varepsilon(\theta^k)\iota(\theta^\ell) + \iota(\theta^k)\varepsilon(\theta^\ell))\mathcal{R}(X_k,X_\ell)^*.
\end{aligned}
$$

Here $\mathcal{R}(\cdot,\cdot)$ is the Riemannian curvature, an endomorphism of the tangent bundle, which acts on forms by the extension of its negative transpose as a derivation. Using the fact that $\varepsilon(\theta^k)$ and $\iota(\theta^\ell)$ anticommute for $k \neq \ell$, and writing the curvature as the endomorphism

$$
\mathcal{R} = \sum_{k<\ell} R_{\cdot k\ell}\,\theta^k \wedge \theta^\ell,
$$

we have the general Weitzenböck formula for differential forms:

$$
(C.11) \qquad dd^* + d^*d = \nabla^*\nabla + \varepsilon(\theta^k)\iota(\theta^\ell)R_{\cdot k\ell}.
$$

Of course, in $(C.11)$ the summation is over all k, ℓ.

Next we compute the action of the correction term on the 2-form $\theta^a \wedge \theta^b$. First, recall that the Ricci curvature is defined by

$$
R^i_k = \sum_m R^i_{mkm},
$$

and the scalar curvature by

$$
R = \sum_m R^m_m.
$$

For convenience we lower all indices. Then the Weyl curvature is given by

$$
W_{ijk\ell} = R_{ijk\ell} - \frac{1}{2}(\delta_{ik}R_{j\ell} - \delta_{i\ell}R_{jk} - \delta_{jk}R_{i\ell} + \delta_{j\ell}R_{ik})
$$

$$
+ \frac{1}{6}(\delta_{ik}\delta_{j\ell} - \delta_{i\ell}\delta_{jk})R,
$$

which can be checked by $\sum_m W_{imkm} = 0$. The Riemannian curvature satisfies the symmetry relations

$$
R_{ijk\ell} = -R_{jik\ell} = -R_{ij\ell k} = R_{k\ell ij}
$$

and the Bianchi identity

$$R_{ijk\ell} + R_{ik\ell j} + R_{i\ell jk} = 0.$$

Finally, writing $\theta^{ij} = \theta^i \wedge \theta^j$, and summing over all indices except a and b, we compute

$$\varepsilon(\theta^k)\iota(\theta^\ell)R_{\cdot k\ell}(\theta^a \wedge \theta^b)$$

$$=\varepsilon(\theta^k)\iota(\theta^\ell)[R_{amk\ell}\theta^{mb} + R_{bmk\ell}\theta^{am}]$$

$$=R_{a\ell k\ell}\theta^{kb} - R_{amkb}\theta^{km} + R_{bmka}\theta^{km} - R_{b\ell k\ell}\theta^{ka}$$

$$=R_{ak}\theta^{kb} + R_{bk}\theta^{ka} - R_{kmab}\theta^{km}$$

$$=R_{ak}\theta^{kb} + R_{bk}\theta^{ka} - W_{kmab}\theta^{kn}$$

$$\qquad - \frac{1}{2}(R_{mb}\theta^{am} - R_{ma}\theta^{bm} - R_{kb}\theta^{ka} + R_{ka}\theta^{kb})$$

$$\qquad + \frac{1}{6}(\theta^{ab} - \theta^{ba})R$$

$$= -W_{kmab}\theta^{km} + \frac{R}{3}\theta^{ab}.$$

The Weyl curvature W acts as an endomorphism by

$$W = \frac{1}{2}\sum_{k,\ell} W_{\cdot k\ell}e^k \wedge e^\ell,$$

just as the full Riemannian curvature does, so

$$(C.12) \qquad\qquad dd^* + d^*d = \nabla^*\nabla - 2W(\cdot) + \frac{R}{3}$$

on 2-forms.

Suppose now that ω_- is an anti-self-dual 2-form. Then

$$(dd*+d*d)\omega_- = -(d*d*\omega_- + *d*d\omega_-)$$

$$= -(-d*d + *d*d)\omega_-$$

$$= (1 - *)d*d\omega_-$$

$$= 2P_-d(-*d*)\omega_-$$

$$= 2d_-d_-^*\omega_-.$$

So $(C.12)$ becomes

$$2d_-d_-^* = \nabla^*\nabla - 2W^-(\cdot) + \frac{R}{3}$$

on anti-self-dual 2-forms.

174

APPENDIX D
THE REMOVABILITY OF SINGULARITIES

Our use of the blown-up manifold in §7 provides a greatly simplified proof of the Removable Singularities Theorem, which also applies to non-self-dual solutions. The situation is that we have a local solution of the full Yang–Mills equations in an open ball B^4 with some metric, not necessarily the flat metric.

THEOREM D.1. *Let D be a solution of the Yang–Mills equations in $B^4-\{0\}$ with $\int_{B^4}|F_D|^2 < \infty$, $D = d+A$, and $A \in L^2_{1,\mathrm{loc}}(B^4-\{0\})$. Then D is gauge equivalent to a connection $\tilde{D}$ which extends smoothly across the singularity to a smooth connection.*

Here $A \in L^2_{1,\mathrm{loc}}(B^4 - \{0\})$ means that $\varphi A \in L^2_1(B^4 - \{0\})$ for every smooth compactly supported function $\varphi \in C_0^\infty(B^4 - \{0\})$.

Our first step is to deform the metric conformally so it is approximately cylindrical, as in §7. Since the Yang–Mills equations and the L^2 norm are conformally invariant, this changes nothing. We work entirely in the cylinder $\{\tau_0 \leq \tau\}$. Set $F = F_D$.

LEMMA D.1. $\lim_{\tau\to\infty}|F(\tau,\theta)| = 0.$

PROOF: Certainly $\lim_{\bar{\tau}\to\infty}\int_{\tau\geq\bar{\tau}}|F(\tau,\theta)|^2 = 0$. However, for balls B_σ over which $\int_{B_\sigma(y_0)}|F(y)|^2dy < \varepsilon$ where ε depends on the geometry, but can be fixed here because the geometry of the cylinder is uniform, we have a regularity theorem (8.3) which says $\sigma^2|F(y_1)|^2 \leq c\int_{B_\sigma(y_0)}|F(y)|^2dy$ if $\mathrm{dist}(y_0,y_1) < \frac{\sigma}{2}$. This is a generalization of the Regularity Theorem (8.3) we gave in §8 to the full Yang–Mills equations.

If we choose $\bar{\tau}$ sufficiently large, then $\int_{\tau\geq\bar{\tau}}|F(\tau,\theta)|^2 < \varepsilon$ and σ can be chosen of uniform size. Hence

$$|F(\bar{\tau},\theta)|^2 \leq \frac{c}{\sigma^2}\int_{\bar{\tau}-1\leq\tau<\bar{\tau}+1}|F(x)|^2dx \to 0.$$

LEMMA D.2. *For any $\gamma < \sqrt{2}$ we can choose $\bar{\tau}$ such that for $\tau \geq \bar{\tau}$,*

$$|F(\tau,\theta)| \leq \max_\theta |F(\bar{\tau},\theta)| \cdot e^{\gamma(\bar{\tau}-\tau)}.$$

175

PROOF: We simply repeat part of the proof of the decay estimates (9.8). By the Weitzenböck formulas, for $\tau \geq \bar{\tau}$

$$|F| + \bar{\gamma}|F| \geq 0$$

for $\bar{\tau}$ chosen so that the metric is close to the cylindrical one and $|F|(\tau, \theta)$ is small for $\tau \geq \bar{\tau}$, $\bar{\gamma} < 2$. By our maximum principle (7.10), for $\bar{\tau} \leq \tau \leq \tau_n$, $\gamma < \sqrt{\bar{\gamma}}$,

$$(D.3) \qquad \begin{aligned} |F(\tau, \theta)| &\leq \max_{\theta} |F(\bar{\tau}, \theta)| e^{\gamma(\bar{\tau} - \tau)} \\ &\quad + \max_{\theta} |F(\tau_n, \theta)| e^{\gamma(\tau - \tau_n)}. \end{aligned}$$

Letting $\tau_n \to \infty$ gives the result.

Next, we mimic our construction of exponential gauge in the neck of the instanton. However, again we can let $\tau_n \to \infty$ and we get a gauge in the entire region $\tau \geq \bar{\tau}$.

LEMMA D.4. *If $\bar{\tau}$ is sufficiently large, there exists a gauge on $\{\tau \geq \bar{\tau}\}$ in which $D = d + A$, where $|A(\tau, \theta)| \leq c \cdot e^{\gamma(\bar{\tau} - \tau)}$.*

See the proof of Lemma 9.38 for this.

Finally, we transfer ourselves back to the ball. In the ball, the estimates (D.3) and (D.4) transform into

$$|A(x)| \leq c|x|^{\gamma - 1},$$

$$|F(x)| \leq c|x|^{\gamma - 2}.$$

This means $A \in L_1^p$ and F is bounded in L^p for $p < \frac{4}{2-\gamma}$. So (8.2), extended to L^p spaces, $p \geq 2$, implies the existence of the Coulomb gauge; i.e., A is gauge equivalent (in fact, through a continuous gauge transformation if $p > 2$) to $\tilde{A} = s^{-1} ds + s^{-1} As$ with $d^* \tilde{A} = 0$, $\tilde{A} \in L_1^p(B^4)$. Now the system

$$d^* \tilde{A} = 0$$

$$d^* d\tilde{A} + d^*(\tilde{A} \wedge \tilde{A}) + (\tilde{A} \lrcorner d\tilde{A}) + (\tilde{A} + (\tilde{A} \wedge \tilde{A})) = 0$$

can be made into the single elliptic equation

$$\Delta \tilde{A} + \text{ lower order terms } = 0.$$

Regularity follows as in Proposition 8.3, and $d + \tilde{A}$ provides the desired extension.

APPENDIX E
TOPOLOGICAL REMARKS

There are a few places in the text where we omitted a topological argument that seemed out of place. Here we take the opportunity to fill in some of those details. These are exercises to our readers versed in topology and may help those without this expertise to get a feel for the ideas involved.

First, the reader may have wondered about our definition of the intersection form. Usually this is taken to be a bilinear form on $H^2(M;\mathbb{Z})/\text{torsion}$, whereas we defined it on $H^2(M;\mathbb{Z})$. The simple explanation is

PROPOSITION E.1. *If* $H_1(M;\mathbb{Z}) = 0$, *then* $H^2(M;\mathbb{Z})$ *is torsion free.*

PROOF: The Universal Coefficient Theorem in cohomology gives

$$H^2(M;\mathbb{Z}) = \text{Hom}(H_2(M;\mathbb{Z}),\mathbb{Z}) \oplus \text{Ext}(H_1(M;\mathbb{Z}),\mathbb{Z}),$$

and since $H_1(M;\mathbb{Z}) = 0$, it follows that $\text{Tor}\, H^2(M;\mathbb{Z}) = 0$.

In the case where $H^2(M;\mathbb{Z})$ contains a torsion element α, say $n\alpha = 0$, then for any $\beta \in H^2(M;\mathbb{Z})$ we have $n(\alpha \smile \beta) = n\alpha \smile \beta = 0$. But $H^4(M;\mathbb{Z})$ is torsion-free, which implies $\alpha \smile \beta = 0$. This proves that the kernel of the bilinear form

$$\smile: H^2(M;\mathbb{Z}) \otimes H^2(M;\mathbb{Z}) \to \mathbb{Z}$$

$$\alpha \otimes \beta \mapsto (\alpha \smile \beta)[M]$$

contains $\text{Tor}\, H^2(M;\mathbb{Z})$, and Poincaré duality over $\mathbb{Q}$—the assertion that $\smile: H^2(M;\mathbb{Q}) \otimes H^2(M;\mathbb{Q}) \to \mathbb{Q}$ is non-degenerate—implies that this kernel is exactly $\text{Tor}\, H^2(M;\mathbb{Z})$. Therefore the induced intersection form

$$\omega : H^2(M;\mathbb{Z})/\text{Tor}\, H^2(M;\mathbb{Z}) \otimes H^2(M;\mathbb{Z})/\text{Tor}\, H^2(M;\mathbb{Z}) \to \mathbb{Z}$$

is nondegenerate.

Our other remarks are of a less trivial nature, and for these we need the notions of *classifying space* and *Eilenberg–MacLane space*. (References for this material are [**BT**], [**H**], [**Sp**], and [**Ste**].) Let G be any topological group. It is a theorem in Topology that there exists a contractible space EG on which G acts freely. Under a mild restriction on G, which is certainly satisfied by all Lie groups, the space EG is a CW complex, unique up to homotopy [**M4**]. The quotient $BG = EG/G$ is the *classifying space* for G, and $G \to EG \to BG$ is a principal fibration.

Examples.

1. $G = \mathbf{Z}$. Then $\mathbf{Z}$ acts freely on $\mathbf{R}$ by translation, $\mathbf{R}$ is contractible, and the quotient $S^1 = \mathbf{R}/\mathbf{Z}$ is $B\mathbf{Z}$.

2. $G = U(1)$. Now, as is more typical, the classifying space is infinite dimensional. The total space of the $U(1)$ bundle $U(1) \to S^{2k+1} \to \mathbf{CP}^k$ is $2k$-connected, and as k increases S^{2k+1} approximates a contractible space. Form the towers

$$(E.2) \qquad \begin{array}{c} S^3 \subset \ S^5 \subset \ S^7 \subset \dots \\ \mathbf{CP}^1 \subset \mathbf{CP}^2 \subset \mathbf{CP}^3 \subset \dots \end{array}$$

by inclusion. The limiting spaces can be defined as CW complexes, which we denote S^∞ and $\mathbf{CP}^\infty$, respectively. Then S^∞ is contractible, and so $U(1) \to S^\infty \to \mathbf{CP}^\infty$ is a universal fibration, whence $BU(1) = \mathbf{CP}^\infty$. Alternatively, let $\mathcal{H}_\mathbf{C}$ be a complex, separable infinite dimensional Hilbert space, $S^\infty \subset \mathcal{H}_\mathbf{C}$ the unit ball, and $\mathbf{CP}^\infty = \mathbf{P}(\mathcal{H}_\mathbf{C})$ its projectivization. These spaces can be identified with the limits in $(E.2)$.

3. $G = SU(2)$. We simply replace the complex numbers in the previous example with the quaternions. Thus

$$S^7 \subset S^{11} \subset S^{15} \subset \dots \subset S^\infty$$
$$\mathbf{HP}^1 \subset \mathbf{HP}^2 \subset \mathbf{HP}^3 \subset \dots \subset \mathbf{HP}^\infty,$$

and the universal fibration is $SU(2) \to S^\infty \to \mathbf{HP}^\infty$, whereby $BSU(2) = \mathbf{HP}^\infty$. Or, $\mathbf{HP}^\infty = \mathbf{P}(\mathcal{H}_\mathbf{H})$, where $\mathcal{H}_\mathbf{H}$ is a separable, quaternionic Hilbert space.

4. $G = SO(3)$. Then $BSO(3)$ is the Grassmannian of oriented 3-planes π in a separable, real Hilbert space $\mathcal{H}_\mathbf{R}$, and $ESO(3)$ is the infinite Stiefel variety whose fiber over π is the set of all oriented orthonormal bases of π.

5. $G = \tilde{\mathcal{G}}$, the group of gauge transformations modulo its center. Then $E\tilde{\mathcal{G}} = \hat{\mathcal{A}}$, the space of irreducible connections, and $B\tilde{\mathcal{G}} = \hat{\mathcal{X}}$ (cf. §3).

The term "classifying space" refers to the fact that BG classifies principal G bundles. Let M be a manifold (more generally, a CW complex) and $f : M \to BG$ a continuous map. We can pull back the G bundle $\xi : EG \to BG$ to obtain a bundle $f^*\xi \to M$. It can be shown that homotopic maps give rise to equivalent bundles, and that the correspondence

$$[M, BG] \leftrightarrow \{\text{equivalence classes of } G \text{ bundles over } M\}$$

is one-to-one. For example, $[M, \mathbf{CP}^\infty]$ classifies $U(1)$ bundles over M, and $[M, \mathbf{HP}^\infty]$ classifies $SU(2)$ bundles.

The *Eilenberg–MacLane spaces* $K(\pi, n)$, which also live in the homotopy category of CW complexes, are the fundamental building blocks of homotopy theory. They are characterized as having a single nontrivial homotopy group:

$$\pi_i(K(\pi, n)) = \begin{cases} \pi & \text{if } i = n, \\ 0 & \text{if } i \neq n. \end{cases}$$

Of course, the group π is required to be abelian if $n \geq 2$. Using the homotopy exact sequences for the universal fibrations in the first two examples above, we can demonstrate that

$$(E.3) \qquad \begin{aligned} K(\mathbf{Z}, 1) &= B\mathbf{Z} = S^1 \\ K(\mathbf{Z}, 2) &= BU(1) = \mathbf{CP}^\infty. \end{aligned}$$

Eilenberg–MacLane spaces are classifying for cohomology. Precisely,

$$(E.4) \qquad H^n(M; \pi) \simeq [M, K(\pi, n)]$$

for any space M. The isomorphism $(E.4)$ is obtained as follows. From the Hurewicz Theorem and the Universal Coefficient Theorem, it is easy to see that $H^n(K(\pi, n); \pi) = \mathrm{Hom}(\pi, \pi)$. Let $\iota : \pi \to \pi$ be the identity map. Associated to any $f : M \to K(\pi, n)$ is the cohomology class $f^*\iota \in H^n(M; \pi)$; this is the correspondence (3.4). In Example 2, we have

$$H^*(\mathbf{CP}^\infty; \mathbf{Z}) = \mathbf{Z}[c_1],$$

where $c_1 \in H^2(\mathbf{CP}^\infty; \mathbf{Z})$ is the universal Chern class, i.e., the first Chern class of the universal $U(1)$ bundle $S^\infty \to \mathbf{CP}^\infty$. For $f : M \to \mathbf{CP}^\infty$, it follows that $f^*c_1 \in H^2(M; \mathbf{Z})$ is the first Chern class of the induced $U(1)$ bundle over M.

Now we can proceed to the topological classification of $U(1)$ and $SU(2)$ bundles, which we stated in §2 and relied on in Proposition 2.11.

THEOREM E.5. *The first Chern class $c_1(\lambda) \in H^2(M; \mathbf{Z})$ classifies $U(1)$ bundles over any CW complex M. The second Chern class $c_2(\eta) \in H^4(M^4; \mathbf{Z})$ classifies $SU(2)$ bundles over any compact oriented 4-manifold M^4.*

PROOF: In the circle bundle case we simply observe that $\mathbf{CP}^\infty$ classifies both $U(1)$ bundles and second cohomology:

$$\{\text{equivalence classes of } U(1) \text{ bundles } \lambda \text{ over } M\}$$

$$\simeq [M, \mathbf{CP}^\infty]$$

$$= [M, K(\mathbf{Z}, 2)]$$

$$= H^2(M; \mathbf{Z}).$$

As observed in the preceeding paragraph, the isomorphism is given by $\lambda \mapsto c_1(\lambda)$.

Next, $SU(2)$ bundles η over M^4 correspond to maps $M^4 \to \mathbf{HP}^\infty$. By the Cellular Approximation Theorem, any map $M^4 \to \mathbf{HP}^\infty$ is homotopic to a map into the 4-skeleton $S^4 = \mathbf{HP}^1 \subset \mathbf{HP}^\infty$. Furthermore, homotopies of such maps can be pushed to homotopies inside the 5-skeleton, which is still S^4, and so

$$[M^4; \mathbf{HP}^\infty] = [M^4, S^4].$$

But for compact oriented M^4, $[M^4, S^4] \simeq \mathbf{Z}$ (cf. Appendix B), and the isomorphism is given by degree. It is not hard to see that this degree is also the second Chern number of the induced $SU(2)$ bundle.

We note that the second Chern number does *not* classify $SU(2)$ bundles over 5-manifolds due to possible torsion in the cohomology.

The results of §10 depend on the classification of $SO(3)$ bundles ξ over a compact oriented 4-manifold $M = M^4$. There are two characteristic classes associated to ξ—the second Stiefel–Whitney class $w_2(\xi) \in H^2(M; \mathbf{Z}_2)$ and the first Pontrjagin class $p_1(\xi) \in H^4(M; \mathbf{Z})$. (There is also a third Stiefel–Whitney class, but it is determined by $w_2(\xi)$.) The group $SU(2)$ can be identified with the simply connected double cover of $SO(3)$, the spin group $\mathrm{Spin}(3)$, and the covering homomorphism is then the adjoint representation $\mathrm{ad} : SU(2) \to SO(3)$. There is a functorially induced map $B(\mathrm{ad}) : BSU(2) \to BSO(3)$ which is the classifying map for the adjoint $SO(3)$ bundle associated to the universal $SU(2)$ bundle $ESU(2) \to BSU(2)$. The characteristic classes of this associated bundle can be computed from the weights of the adjoint representation [**BH**]. Thus

$$(E.6) \qquad \begin{aligned} B(\mathrm{ad})^* w_2 &= 0 \\ B(\mathrm{ad})^* p_1 &= -4c_2, \end{aligned}$$

180

where $w_2 \in H^2(BSO(3); \mathbb{Z}_2)$, $p_1 \in H^4(BSO(3); \mathbb{Z})$, and $c_2 \in H^4(BSU(2), \mathbb{Z})$ are the appropriate universal characteristic classes. Note that the first equation follows simply from $H^2(BSU(2); \mathbb{Z}) = 0$, and reflects the fact that $w_2(\xi)$ is the obstruction to a spin structure on ξ, i.e., to a lift $\tilde{f} : M \to BSU(2)$ of the classifying map $f : M \to BSO(3)$ of ξ. Finally, the universal class $w_2 \in H^2(BSO(3); \mathbb{Z}_2)$ represents a map $w_2 : BSO(3) \to K(\mathbb{Z}_2, 2)$ by $(E.4)$, and since any map in homotopy theory is a fibration [**BT**, p. 249], we obtain a fibering $F \to BSO(3) \to K(\mathbb{Z}_2, 2)$ for some space F. Now $w_2 \circ B(\mathrm{ad}) = 0$ by $(E.6)$, so $B(\mathrm{ad})$ is homotopic to a map $\theta : BSU(2) \to F$. From the usual exact sequences we conclude that θ induces an isomorphism of homotopy groups, and then the Hurewicz Theorem implies that θ is a homotopy equivalence. So up to homotopy equivalence we have the fibration

$$(E.7) \qquad \begin{array}{c} BSU(2) \xrightarrow{B(\mathrm{ad})} BSO(3) \\ \downarrow w_2 \\ K(\mathbb{Z}_2, 2). \end{array}$$

With these preliminaries, we can state

THEOREM E.8. *Let ξ_1, ξ_2 be $SO(3)$ bundles over a compact oriented 4-manifold M such that $w_2(\xi_1) = w_2(\xi_2)$ and $p_1(\xi_1) = p_1(\xi_2)$. Then ξ_1 and ξ_2 are topologically equivalent.*

Strictly speaking, this is not a *classification* since we do not specify which pairs w_2, p_1 occur. However, in §10 we only need the *characterization* given here. The idea of the proof is clear. Since $w_2(\xi_1) = w_2(\xi_2)$, the "difference" "$\xi_2 - \xi_1$" lifts to an $SU(2)$ bundle whose Chern class $-\frac{1}{4}(p_1(\xi_2) - p_1(\xi_1))$ vanishes. Then $\xi_1 = \xi_2$ by the previous theorem. Of course, there is no way to make sense of "$\xi_2 - \xi_1$" as an $SO(3)$ bundle. The argument we give, kindly supplied by Richard Lashof, puts this intuitive idea in a rigorous context.

PROOF: Let $f_1, f_2 : M \to BSO(3)$ be classifying maps for ξ_1, ξ_2 respectively; we will prove that f_1 is homotopic to f_2. First, $w_2 \circ f_1$ and $w_2 \circ f_2$ are homotopic maps $M \to K(\mathbb{Z}_2, 2)$, since $w_2(\xi_1) = w_2(\xi_2)$. Denote the homotopy by $h : M \times [0,1] \to K(\mathbb{Z}, 2)$. We try to lift h to a homotopy $\tilde{h} : M \times [0,1] \to BSO(3)$ step-by-step using a cell structure on M. Suppose that $\tilde{h}_{i-1}$ can be defined on the $(i-1)$-skeleton of M (cross $[0,1]$), and consider an i-cell $\Delta^i \subseteq M$. The bundle $(E.7)$ is trivial over

$h(\Delta^i \times [0,1])$, so that f_1 and f_2 together with the lift $\tilde{h}_{i-1}$ determines a map $S^i \to BSU(2)$, that is, an element of $\pi_i(BSU(2))$.

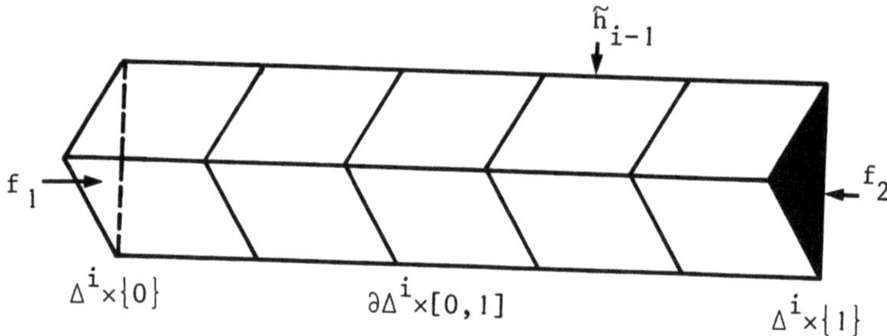

Then $\tilde{h}_{i-1}$ can be extended over Δ^i if and only if this is the zero element. Since $\pi_i(BSU(2)) = \pi_{i-1}(SU(2))$ vanishes for $i \le 3$, it follows by induction that there is a lift $\tilde{h}_3$ over the 3-skeleton of M. To simplify matters, assume that the cell decomposition of M contains only one 4 cell Δ^4 (this can always be arranged). Also, we can assume (via the homotopy $\tilde{h}_3$) that $f_1 = f_2$ on the 3-skeleton. Now by pinching the middle of the cell Δ^4 we obtain a map $\sigma : M \to M \vee S^4$. Furthermore, the induced map on cohomology $\sigma^* : H^*(M \vee S^4) = H^*(M) \oplus H^*(S^4) \to H^*(M)$ is simply addition. We can write $f_2 = (f_1 \vee \gamma) \circ \sigma$ for

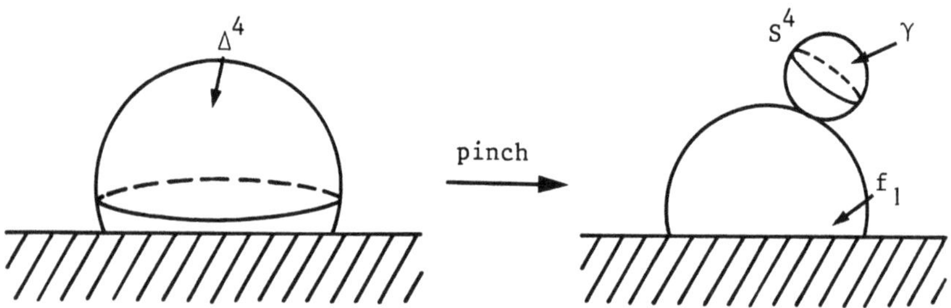

some map $\gamma : S^4 \to BSO(3)$ since $f_1 = f_2$ off of Δ^4. There is a lift $\tilde{\gamma} : S^4 \to BSU(2)$, because $H^2(S^4; \mathbb{Z}_2) = 0$. (In other words, every $SO(3)$ bundle over S^4 carries a spin structure.) Now a simple calculation using $(E.6)$ yields

$$\tilde{\gamma}^*(c_2) = -\frac{1}{4}(p_1(\xi_2) - p_1(\xi_1)) = 0.$$

The classification of $SU(2)$ bundles $(E.5)$ implies that $\tilde{\gamma}$, whence γ, is homotopically trivial. Therefore, f_1 is homotopic to f_2 as claimed. Note that

γ makes precise the difference "$\xi_2 - \xi_1$" discussed before the proof.

The notion of a Eilenberg–MacLane space, in particular the smooth manifold $\mathbf{CP}^\infty$, plays a role in proving a fact that we used implicitly in §1 and in Appendix B.

PROPOSITION E.9. *Let M be a smooth compact oriented manifold of dimension n, and $z \in H_{n-2}(M; \mathbf{Z})$ a codimension two homology class. Then z can be represented by a smooth oriented submanifold of M.*

PROOF: By Poincaré duality the homology class z corresponds to a cohomology class $z^* \in H^2(M; \mathbf{Z})$. Now $(E.4)$ implies that z^* is represented by a map $f : M \to K(\mathbf{Z}, 2) = \mathbf{CP}^\infty$. Recalling the "telescoping construction" $(E.2)$ for $\mathbf{CP}^\infty$, and using the compactness of M, we conclude that the image of f lies in some finite dimensional $\mathbf{CP}^N$. By perturbing f in its homotopy class we can assume that f is transverse to $\mathbf{CP}^{N-1} \subset \mathbf{CP}^N$. Then $f^{-1}(\mathbf{CP}^{N-1})$ is the desired smooth representative of z.

BIBLIOGRAPHY

[**Ar**] N. Aronszajn, *A unique continuation theorem for solutions of elliptic partial differential equations or inequalities of the second order*, J. Math. Pures Appl. (9) **36** (1957), 235–249.

[**A1**] M. F. Atiyah, "*K*-Theory," W. A. Benjamin, Inc., 1967.

[**A2**] __________, "Geometry of Yang–Mills Fields," Lezioni Fermiane, Accademia Nazionale dei Lincei Scuola Normale Superione, Pisa, 1979.

[**AB**] M. F. Atiyah and R. Bott, *The Yang–Mills equations over Riemann surfaces*, Proc. R. Soc. London A **308** (1982), 523–615.

[**ABS**] M. F. Atiyah, R. Bott and A. Shapiro, *Clifford modules*, Topology **3**(1964), Suppl. 1, 3–38.

[**AHS**] M. F. Atiyah, N. Hitchin and I. M. Singer, *Self-duality in four-dimensional Riemannian geometry*, Proc. R. Soc. London A **362** (1978), 425–461.

[**APS**] M. F. Atiyah, V. K. Patodi and I. M. Singer, *Spectral asymmetry and Riemannian geometry* I, Math. Proc. Comb. Phil. Soc. **77** (1975), 43–69.

[**AS**] M. F. Atiyah and I. M. Singer, *The index of elliptic operators:* IV, Annals of Mathematics **93** (1971), 119–138.

[**Au**] T. Aubin, "Nonlinear Analysis on Manifolds, Monge–Ampere Equations," Springer–Verlag, 1982.

[**BPST**] A. Belavin, A. Polyakov, A. Schwartz and Y. Tyupkin, Phys. Let. **59B** (1975), 85.

[**B**] D. Bleeker, "Gauge Theory and Variational Principles," Addison Wesley, London, 1981.

[**Bo**] S. Bochner, *Vector fields and Ricci curvature*, Bull. Amer. Math. Soc. **52** (1946), 776–797.

[**BH**] A. Borel and F. Hirzebruch, *Characteristic classes and homogeneous spaces*, Part I, American Journal of Mathematics **80** (1958), 458–538.

[**BT**] R. Bott and L. Tu, "Differential Forms in Algebraic Topology," Springer–Verlag, 1982.

[**Bou**] J. P. Bourguignon, *Formules de Weitzenböck en dimension* 4, in "Géometrie Riemannienne de dimension 4," CEDIC, Paris, 1981.

[**BL**] J. P. Bourguignon and H. B. Lawson, Jr., *Yang Mills Theory: Its physical origins and differential geometric aspects*, in "Seminar on Differential Geometry," S. T. Tau, ed., Annals of Mathematics Studies Number 102, Princeton University Press, 1982, 395–422.

[C] A. Casson, *Three lectures on new infinite constructions in 4-dimensional manifolds*, Notes by Lucian Guillou, Prépublications Orsay **81 T 06**.

[Ce] J. Cerf, "Sur les Difféomorphismes de la Sphère de Dimension Trois $(\Gamma_4 = 0)$," Lecture Notes in Mathematics **53**, Springer–Verlag, 1968.

[**CE**] J. Cheeger and D. Ebin, "Comparison Theorems in Riemannian Geometry," North Holland Mathematical Library, Vol. 9, 1975.

[**CS**] J. H. Conway and N. J. A. Sloane, *On the enumeration of lattices of determinant one*, J. Number Theory **15** (1982), 83–94.

[**CH**] R. Courant and D. Hilbert, "Methods of Mathematical Physics," Interscience, 1953.

[**DW**] A. Dold and H. Whitney, *Classification of oriented sphere bundles over a 4-complex*, Annals of Mathematics **69** (1959), 667–677.

[D] S. K. Donaldson, *An application of gauge theory to the topology of 4-manifolds*, J. Diff. Geo. **18** (1983), 269–316.

[**Do**] L. Dornhoff, "Group Representation Theory, Part A," Pure and Applied Mathematics, Marcel Dekker, Inc., 1971.

[**DV**] A. Douady and J. L. Verdier, *Les équations de Yang–Mills*, Astérisque **71–72**, Société Mathématique de France, 1980.

[**FS**] R. Fintushel and R. Stern, *New invariants for homology 3-spheres*, to appear.

[**Fr**] M. Freedman, *The topology of four-dimensional manifolds*, J. Diff. Geo. **17** (1982), 357–454.

[**Fr2**] ——————, *There is no room to spare in four dimensional space*, AMS Notices **31** (1984), 3–6.

[**FK**] M. Freedman and R. Kirby, *A geometric proof of Rohlin's theorem*, Proc. Symp. Pure Math. **32:2** (1978), 85–98.

[F] A. Friedman, "Partial Differential Equations," Holt, New York, 1969.

[**GT**] D. Gilbarg and N. S. Trudinger, "Elliptic Partial Differential Equations of Second Order," Springer–Verlag, 1977.

[G] R. Gompf, *Three exotic $\mathbf{R}^4$'s and other anomalies*, J. Diff. Geo. **18** (1983), 317–328.

[**GW**] R. E. Green and H. Wu, "Function Theory on Manifolds Which Possess a Pole," Lecture Notes in Mathematics **699**, Springer-Verlag,

1979.

[**He**] J. Hempel, "Three-Manifolds," Annals of Mathematics Studies, Number **86**, Princeton University Press, 1967.

['] Gerard 't Hooft, *Gauge theories of the forces between elementary particles*, Scientific American **243**(June 1980), 104–138.

[**H**] D. Husemoller, "Fibre Bundles," Second Edition, Springer–Verlag, 1966.

[**HM**] D. Husemoller and J. Milnor, "Symmetric Bilinear Forms," Springer–Verlag, 1973.

[**JNR**] R. Jackiw, C. Nohl and C. Rebbi, *Conformal properties of pseudoparticle configurations*, Phys. Rev. D. **15** (1977), 1642–1646.

[**JT**] A. Jaffe and C. Taubes, "Vortices and Monopoles," Progress in Physics 2, Birkhäuser, 1980.

[**KM1**] M. Kervaire and J. Milnor, *Bernoulli numbers, homotopy groups, and a theorem of Rohlin*, Proc. Internat. Congr. Math. Edingburgh (1958), 454–458.

[**KM2**] ______________, *Groups of homotopy spheres* I, Annals of Mathematics **77** (1963), 504–537.

[**KS**] R. Kirby and L. Siebenmann, "Foundational Essays on Topological Manifolds, Smoothings, and Triangulations," Annals of Mathematics Studies, Number 88, Princeton University Press, Princeton, NJ, 1977.

[**KN1**] S. Kobayashi and K. Nomizu, "Foundations of Differential Geometry," Vol. 1, Wiley–Interscience, 1963.

[**KN2**] ______________, "Foundations of Differential Geometry," Vol. 2, Wiley–Interscience, 1969.

[**K**] M. Kuranishi, *New proof of the existence of locally complete families of complex structures*, Proceedings of the Conference on Complex Analysis, A. Aeppli et al, eds., Springer–Verlag (1965), 142–154.

[**L**] A. Lichnerowicz, *Spineurs harmoniques*, C.R. Acad. Sci. Paris Ser. A-B **257** (1963), 7–9.

[**Ma**] J. Marsden, "Applications of Global Analysis in Mathematical Physics," Publish or Perish, 1974.

[**May**] M. Mayer, in "Fiber–Bundle Techniques in Gauge Theory," Lecture Notes in Physics, Vol. 67, Springer–Verlag, Berlin–Heidelberg–New York, 1977.

[**MSY**] W. Meeks III, L. Simon and S. T. Yau, *Embedded minimal surfaces, exotic spheres, and manifolds with positive Ricci curvature, An-*

nals of Mathematics **116** (1982), 621–659.

[**MY1**] W. Meeks III and S. T. Yau, *The classical Plateau problem and the topology of three dimensional manifolds*, Topology **21** (1982), 409–442.

[**MY2**] ______________, *Topology of three dimensional manifolds and the embedding problems in minimal surface theory*, Annals of Mathematics **112** (1980), 441–484.

[**M1**] J. Milnor, *On manifolds homeomorphic to the 7-sphere*, Annals of Mathematics **64** (1956), 399–405.

[**M2**] ________, *On simply connected 4-manifolds*, Symposium Internacional Topologia Algebraica, Mexico (1958), 122–128.

[**M3**] ________, "Topology from the Differentiable Viewpoint," University Press of Virginia, 1976.

[**M4**] ________, *Construction of universal bundles* II, Annals of Mathematics **63** (1956), 430–436.

[**M5**] ________, "Morse Theory," Annals of Mathematics Studies, Number 51, Princeton Univesity Press, Princeton, NJ, 1963.

[**MS**] J. Milnor and J. Stasheff, "Characteristic Classes," Annals of Mathematics Studies, Number 76, Princeton University Press, Princeton, NJ, 1974.

[**MTW**] C. W. Misner, K. S. Thorne and J. A. Wheeler, "Gravitation," W. H. Freeman and Company, 1973.

[**Mo**] E. Moise, *Affine Structure in 3-manifolds* I, II, III, IV, V, Annals of Mathematics **54**(1951), 506–533; **55**(1952), 172–176; **55**(1952), 203–214; **55**(1952), 215–222; **56**(1952), 96–114.

[**Mor**] C. B. Morrey, Jr., "Multiple Integrals in the Calculus of Variations," Springer–Verlag, 1966.

[**P1**] R. S. Palais, "Foundations of Global Non-Linear Analysis," W. A. Benjamin, Inc., 1968.

[**P2**] ________, "Seminar on the Atiyah–Singer Index Theorem," Annals of Mathematics Studies, Number 57, Princeton University Press, Princeton, NJ, 1965.

[**Pa**] T. Parker, *Gauge theories on four dimensional Riemannian manifolds*, Comm. Math. Phys. **85** (1982), 1–40.

[**Pe**] F. P. Peterson, *Some remarks on Chern classes*, Annals of Mathematics **69** (1959), 414–420.

[**Q**] F. Quinn, *Ends* III, J. Diff. Geo. **17** (1982), 503–521.

[**R**] V. A. Rohlin, *New results in the theory of 4-dimensional manifolds* (Russian), Dok. Akad. Nauk, USSR **84** (1952), 221–224.

[**SU**] J. Sacks and K. Uhlenbeck, *The existence of minimal 2-spheres*, Annals of Mathematics **113** (1981), 1–24.

[**Se**] S. Sedlacek, *A direct method for minimizing the Yang–Mills functional on 4-manifolds*, Comm. Math. Phys. **86** (1982), 515–527.

[**S**] I. M. Singer, *Some remarks on the Gribov ambiguity*, Comm. Math. Phys. **60** (1978), 7–12.

[**ST**] I. M. Singer and J. A. Thorpe, *The curvature of 4-dimensional Einstein spaces*, in "Global Analysis, Papers in honor of K. Kodaira," (ed. D. Spencer and S. Iyanaga), Princeton University Press (1969), 335–365.

[**Sm**] S. Smale, *An infinite dimensional version of Sard's theorem*, Amer. J. Math. **87** (1968), 861–866.

[**Sp**] E. Spanier, "Algebraic Topology," McGraw Hill, Inc., 1966.

[**St**] J. Stallings, *The piecewise-linear structure of Euclidean space*, Proc. Cambridge Philos. Soc. **58** (1962), 481–488.

[**St2**] _________, "Group Theory and Three-Dimensional Manifolds," Yale Mathematical Monographs, No. 4, Yale University Press, 1971.

[**Ste**] N. Steenrod, "The Topology of Fibre Bundles," Princeton University Press, Princeton, NJ, 1951.

[**Sn**] R. Stern, *Instantons and the topology of 4-manifolds*, The Mathematical Intelligencer **5:3** ((1983)), 39–44.

[**Sto**] R. E. Stong, "Notes on Cobordism Theory," Princeton University Press, 1968.

[**T**] C. H. Taubes, *Self-dual Yang–Mills connections on non-self-dual 4-manifolds*, J. Diff. Geo. **17** (1982), 139–170.

[**T2**] _________, *Self-dual connections on 4-manifolds with indefinite intersection matrix*, preprint.

[**Tr**] F. Trèves, "Basic Linear Partial Differential Equations," Academic Press, New York, 1975.

[**U1**] K. Uhlenbeck, *Connections with L^p bounds on curvature*, Comm. Math. Phys. **83** (1982), 31–42.

[**U2**] _________, *Removable singularities in Yang–Mills fields*, Comm. Math. Phys. **83** (1982), 11–29.

[**V**] V. S. Varadarajan, "Lie Groups, Lie Algebras, and their Representations," Prentice Hall, 1974.

[**Wa**] F. Warner, "Foundations of Differentiable Manifolds and Lie Groups," Springer–Verlag, 1983.

[**W**] J. H. C. Whitehead, *On simply connected 4-dimensional polyhedra*, Comment. Math. Helv. **22** (1949), 48–92.